# MANAGING A REAL-TIME MASSIVELY-PARALLEL NEURAL ARCHITECTURE

A THESIS SUBMITTED TO THE UNIVERSITY OF MANCHESTER
FOR THE DEGREE OF DOCTOR OF PHILOSOPHY
IN THE FACULTY OF ENGINEERING AND PHYSICAL SCIENCES

2012

By
James Cameron Patterson
School of Computer Science

# Contents

**Abstract**     11

**Declaration**     12

**Copyright**     13

**Acknowledgements**     14

**The Author**     15

**1 Introduction**     16
    1.1 Research Motivation . . . . . . . . . . . . . . . . . . . . . . . . 17
       1.1.1 Not Chips, but *Cores* with Everything . . . . . . . . . . . 17
       1.1.2 Large System Management . . . . . . . . . . . . . . . . . 20
       1.1.3 Neural Networks on Parallel Computers . . . . . . . . . . 20
       1.1.4 Monitoring the SpiNNaker Neural System . . . . . . . . . 21
    1.2 Contributions . . . . . . . . . . . . . . . . . . . . . . . . . . . . 23
    1.3 Publications . . . . . . . . . . . . . . . . . . . . . . . . . . . . . 24
    1.4 Thesis Overview . . . . . . . . . . . . . . . . . . . . . . . . . . 26

**2 Neural Computing**     29
    2.1 Biology . . . . . . . . . . . . . . . . . . . . . . . . . . . . . . . 30
    2.2 Neural Network Modelling . . . . . . . . . . . . . . . . . . . . . 33
       2.2.1 First Generation Artificial Neural Networks (ANNs) . . . . 34
       2.2.2 Second Generation ANNs . . . . . . . . . . . . . . . . . . 34
       2.2.3 Third Generation ANNs . . . . . . . . . . . . . . . . . . . 36
    2.3 Systems for Neural Network Modelling . . . . . . . . . . . . . . 39
       2.3.1 General Purpose Hardware . . . . . . . . . . . . . . . . . 40
       2.3.2 Specialised Hardware Platforms . . . . . . . . . . . . . . 45

|  |  | 2.3.3 | The Centre Ground . . . . . . . . . . . . . . . . . . . . | 48 |

# 3 The SpiNNaker Neural Architecture — 50

- 3.1 Architectural Requirements . . . . . . . . . . . . . . . . . . . . . . 51
- 3.2 System Architecture . . . . . . . . . . . . . . . . . . . . . . . . . . 51
- 3.3 SpiNNaker Systems . . . . . . . . . . . . . . . . . . . . . . . . . . 53
- 3.4 SpiNNaker Communications . . . . . . . . . . . . . . . . . . . . . 55
  - 3.4.1 Operation . . . . . . . . . . . . . . . . . . . . . . . . . . . 55
  - 3.4.2 SpiNNaker NoCs and the Router . . . . . . . . . . . . . . . 56
  - 3.4.3 Packet Formats . . . . . . . . . . . . . . . . . . . . . . . . 57
  - 3.4.4 Layered Networking . . . . . . . . . . . . . . . . . . . . . 60
  - 3.4.5 Ex-System, Ethernet . . . . . . . . . . . . . . . . . . . . . 61
  - 3.4.6 Internet Encapsulation . . . . . . . . . . . . . . . . . . . . 62
  - 3.4.7 SpiNNaker Datagram Protocol (SDP) . . . . . . . . . . . . 62
- 3.5 Software on SpiNNaker . . . . . . . . . . . . . . . . . . . . . . . . 63
- 3.6 Summary . . . . . . . . . . . . . . . . . . . . . . . . . . . . . . . 66

# 4 Bootstrapping SpiNNaker — 67

- 4.1 Node-Boot . . . . . . . . . . . . . . . . . . . . . . . . . . . . . . . 68
  - 4.1.1 First Steps . . . . . . . . . . . . . . . . . . . . . . . . . . 71
  - 4.1.2 Monitor Processor Arbitration . . . . . . . . . . . . . . . . 72
  - 4.1.3 Chip-Level POST and Initialisation . . . . . . . . . . . . . 73
  - 4.1.4 Processor-Level Initialisation . . . . . . . . . . . . . . . . 76
  - 4.1.5 The Main Loop . . . . . . . . . . . . . . . . . . . . . . . . 77
- 4.2 Loading the System-Boot Image . . . . . . . . . . . . . . . . . . . 77
  - 4.2.1 Ethernet Flood-Fill . . . . . . . . . . . . . . . . . . . . . . 78
  - 4.2.2 Inter-Chip Flood-Fill . . . . . . . . . . . . . . . . . . . . . 79
- 4.3 IVB – the ITCM Validation Block . . . . . . . . . . . . . . . . . . 82
- 4.4 DHCP Node-Boot Image . . . . . . . . . . . . . . . . . . . . . . . 83
- 4.5 Results from Node-Boot Operation . . . . . . . . . . . . . . . . . . 87
  - 4.5.1 Fault Detection and Isolation . . . . . . . . . . . . . . . . . 87
  - 4.5.2 Response Time and Data Rate . . . . . . . . . . . . . . . . 87
  - 4.5.3 Time taken to Flood-Fill a System-Boot Image . . . . . . . 91
- 4.6 Application Loading . . . . . . . . . . . . . . . . . . . . . . . . . 93
- 4.7 Summary and Contributions . . . . . . . . . . . . . . . . . . . . . 93

# 5 Imaging Neural Networks . . . . . . . . . . . . . . . . . . . . . . . . . . . . . 96
## 5.1 'Wetware' Monitoring . . . . . . . . . . . . . . . . . . . . . . . . . . . 96
### 5.1.1 In-Vivo Structural Imaging . . . . . . . . . . . . . . . . . 98
### 5.1.2 In-Vivo Functional Activity Imaging . . . . . . . . . . . 100
### 5.1.3 Single-Unit Monitoring . . . . . . . . . . . . . . . . . . . 104
### 5.1.4 Biological Imaging Software . . . . . . . . . . . . . . . . 105
## 5.2 Artificial Neural Network Monitoring . . . . . . . . . . . . . . . . . 106
### 5.2.1 Non-Spiking Neural Networks . . . . . . . . . . . . . . . 107
### 5.2.2 Spiking Neural Networks . . . . . . . . . . . . . . . . . . 110
## 5.3 Summary . . . . . . . . . . . . . . . . . . . . . . . . . . . . . . . . . 118

# 6 Visualising Neural Networks on SpiNNaker . . . . . . . . . . . . . . . . . 119
## 6.1 Principles of Visualisation . . . . . . . . . . . . . . . . . . . . . . . . 120
## 6.2 It's All Just Data . . . . . . . . . . . . . . . . . . . . . . . . . . . . . 121
## 6.3 Visualisation Targets . . . . . . . . . . . . . . . . . . . . . . . . . . . 121
## 6.4 Implementing SpiNNaker's Visualiser . . . . . . . . . . . . . . . . . . 124
### 6.4.1 Execution Environment . . . . . . . . . . . . . . . . . . . 124
### 6.4.2 Packet Decode . . . . . . . . . . . . . . . . . . . . . . . . 124
### 6.4.3 Visualisation Interface . . . . . . . . . . . . . . . . . . . . 127
### 6.4.4 Mapping . . . . . . . . . . . . . . . . . . . . . . . . . . . 129
### 6.4.5 SpiNNaker Neural Software and Accessing its Data . . . . 130
## 6.5 Results . . . . . . . . . . . . . . . . . . . . . . . . . . . . . . . . . . 131
### 6.5.1 Neural Dynamics . . . . . . . . . . . . . . . . . . . . . . . 131
### 6.5.2 Spike Activity . . . . . . . . . . . . . . . . . . . . . . . . 133
### 6.5.3 Aggregated Information . . . . . . . . . . . . . . . . . . . 134
### 6.5.4 Interaction . . . . . . . . . . . . . . . . . . . . . . . . . . 135
### 6.5.5 V1 Modelling, Multi-View Simulation . . . . . . . . . . . 135
### 6.5.6 Manipulating Visualiser Data . . . . . . . . . . . . . . . . 137
### 6.5.7 Capacity of the Visualiser . . . . . . . . . . . . . . . . . . 137
## 6.6 Non-Neural Plotting . . . . . . . . . . . . . . . . . . . . . . . . . . . 138
## 6.7 Summary and Contributions . . . . . . . . . . . . . . . . . . . . . . . 141

# 7 Managing Large Network Attached Systems . . . . . . . . . . . . . . . . . 143
## 7.1 Principles of System Management . . . . . . . . . . . . . . . . . . . . 143
### 7.1.1 Fault Management . . . . . . . . . . . . . . . . . . . . . . 144
### 7.1.2 Configuration Management . . . . . . . . . . . . . . . . . 144

|       | 7.1.3 | Accounting Management | 145 |
|---|---|---|---|

       7.1.3  Accounting Management . . . . . . . . . . . . . . . . . 145
       7.1.4  Performance Management . . . . . . . . . . . . . . . . 145
       7.1.5  Security Management . . . . . . . . . . . . . . . . . . 146
  7.2  Managing Large Systems . . . . . . . . . . . . . . . . . . . . 146
       7.2.1  Remote Monitoring . . . . . . . . . . . . . . . . . . . 146
       7.2.2  Hardware Management . . . . . . . . . . . . . . . . . 147
       7.2.3  Systems Management Software . . . . . . . . . . . . . 150
       7.2.4  Protocol and Schema Standards . . . . . . . . . . . . 150
       7.2.5  SNMP: A Walk Through . . . . . . . . . . . . . . . . 153
       7.2.6  The Management Information Base (MIB) . . . . . . . 155
       7.2.7  Previous Large System Research with SNMP . . . . . . . 157
  7.3  Summary . . . . . . . . . . . . . . . . . . . . . . . . . . . . 158

## 8 SpiNNaker Management Framework     160

  8.1  SpiNNaker – a Memory-Mapped Architecture . . . . . . . . . . 160
  8.2  A Protocol Translator – SpiNNmate . . . . . . . . . . . . . . . 161
       8.2.1  Primitive Operations on SpiNNaker . . . . . . . . . . . 163
       8.2.2  SpiNNmate Structure . . . . . . . . . . . . . . . . . . 163
       8.2.3  Protocol Modules . . . . . . . . . . . . . . . . . . . . 166
  8.3  Memory and Communications . . . . . . . . . . . . . . . . . 169
       8.3.1  Memory Operations – When to use DMA . . . . . . . . 169
       8.3.2  Memory Performance Optimisation . . . . . . . . . . . 171
       8.3.3  Communications Bandwidth . . . . . . . . . . . . . . 175
  8.4  Management Framework Results . . . . . . . . . . . . . . . . 176
       8.4.1  Memory Operations – the Performance of SpiNNmate . . . . 176
       8.4.2  A Real-Time SNMP Temperature Plotter . . . . . . . . . 178
       8.4.3  Processor Utilisation Monitor . . . . . . . . . . . . . . 179
       8.4.4  A Network Packet Counter . . . . . . . . . . . . . . . 180
       8.4.5  Long Term Monitoring . . . . . . . . . . . . . . . . . 180
       8.4.6  Alerting using Nagios . . . . . . . . . . . . . . . . . . 182
  8.5  Monitoring a SpiNNaker Machine in Production . . . . . . . . . 183
  8.6  Summary and Contributions . . . . . . . . . . . . . . . . . . 184

## 9 Discussion and Conclusions     186

  9.1  Bootstrapping SpiNNaker . . . . . . . . . . . . . . . . . . . . 187
       9.1.1  Primary Targets . . . . . . . . . . . . . . . . . . . . . 187

|     |       |                                              |     |
|-----|-------|----------------------------------------------|-----|
|     | 9.1.2 | Time Taken to Boot                           | 188 |
|     | 9.1.3 | Power-On Self-Test                           | 188 |
|     | 9.1.4 | ITCM Validation Block (IVB)                  | 189 |
|     | 9.1.5 | DHCP Node-Boot Image                         | 189 |
|     | 9.1.6 | Future Work                                  | 190 |
| 9.2 | Visualising Neural Networks on SpiNNaker     |                  | 192 |
|     | 9.2.1 | Modularity                                   | 192 |
|     | 9.2.2 | Modalities                                   | 192 |
|     | 9.2.3 | Interaction                                  | 193 |
|     | 9.2.4 | Aggregation                                  | 194 |
|     | 9.2.5 | Visualiser Limitations and Future Work       | 195 |
| 9.3 | SpiNNaker Management Framework               |                  | 196 |
|     | 9.3.1 | Protocol Translation                         | 196 |
|     | 9.3.2 | SpiNNmate Performance                        | 198 |
|     | 9.3.3 | Future Work                                  | 199 |
| 9.4 | General SpiNNaker Observations               |                  | 201 |
|     | 9.4.1 | Interconnecting Larger Machines              | 202 |
|     | 9.4.2 | Improving Load and Save Times                | 202 |
|     | 9.4.3 | Further Works in Software                    | 203 |
| 9.5 | Summary                                      |                  | 203 |

**A  Expansion of Abbreviations**     **204**

**Bibliography**     **209**

Word Count: 55371

# List of Tables

| | | |
|---|---|---:|
| 4.1 | System RAM register allocations following Node-Boot | 72 |
| 4.2 | Power-on self-tests and actions | 73 |
| 4.3 | ITCM validation block structure. | 83 |
| 6.1 | Spike-rates found in fig. 6.10 simulation | 138 |
| 8.1 | Loading SDRAM: SpiNNmate's performance | 178 |

# List of Figures

| | | |
|---|---|---|
| 1.1 | SpiNNaker torus | 20 |
| 1.2 | The system management time-line of SpiNNaker. | 22 |
| 2.1 | The human brain. | 30 |
| 2.2 | A six-layer human cortical column | 31 |
| 2.3 | Simplified neuronal structure. | 32 |
| 2.4 | Threshold Logic Unit (TLU) | 34 |
| 2.5 | Second generation Artificial Neural Network | 35 |
| 2.6 | A three-layer Multi-Layer Perceptron (MLP). | 36 |
| 2.7 | Izhikevich spiking model behaviour | 37 |
| 2.8 | A third generation ANN | 38 |
| 3.1 | Photographs of the SpiNNaker chip | 52 |
| 3.2 | SpiNNaker chip schematic | 53 |
| 3.3 | SpiNNaker machine tiled array | 54 |
| 3.4 | SpiNNaker packet formats. | 57 |
| 3.5 | Multicast and Point to Point packet flow and distribution trees. | 58 |
| 3.6 | Nearest Neighbour and Fixed Route packet flow and distribution trees. | 60 |
| 3.7 | SDP Framing Format | 61 |
| 3.8 | Communication layering diagram | 63 |
| 3.9 | Partition And Configuration MANager (PACMAN) | 64 |
| 4.1 | Management time-line: boot | 67 |
| 4.2 | SpiNNaker boot and application loading sequence. | 68 |
| 4.3 | Labelled SpiNNaker chip plot | 69 |
| 4.4 | Node-Boot flowchart (1 of 2) | 70 |
| 4.5 | Node-Boot flowchart (2 of 2) | 74 |
| 4.6 | Node-Boot SDRAM testing procedure | 75 |
| 4.7 | Basic SpiNNaker Ethernet framing / packet format | 78 |

| | | |
|---|---|---|
| 4.8 | Ethernet flood-fill packet types | 79 |
| 4.9 | System-Boot flood-fill 'waves' | 80 |
| 4.10 | Node-Boot to System-Boot state diagram | 81 |
| 4.11 | DHCP address lease assignment process | 85 |
| 4.12 | Auto-discovery of SpiNNaker board on the network | 86 |
| 4.13 | A third generation SpiNNaker 4-chip test board | 87 |
| 4.14 | ICMP echo request – Node-Boot software performance. | 89 |
| 4.15 | Flood-filling the System-Boot image | 92 |
| 4.16 | Internal flood-fill packet types | 95 |
| 5.1 | Structural representations of the human brain. | 97 |
| 5.2 | CAT scanner and output example | 98 |
| 5.3 | Structural MRI image of author | 100 |
| 5.4 | Raw and augmented use of PET data | 101 |
| 5.5 | A functional MRI activity scan | 102 |
| 5.6 | A SQUID skullcap | 102 |
| 5.7 | EEG skullcap and trace | 103 |
| 5.8 | Micro-array of inter-cranial electrodes | 104 |
| 5.9 | Hinton diagram | 108 |
| 5.10 | Bond and Hyperplane static representations. | 108 |
| 5.11 | Trajectory diagram | 109 |
| 5.12 | Raster plot of spike trains | 111 |
| 5.13 | Pattern spotting within raster plots | 112 |
| 5.14 | Nengo output | 115 |
| 6.1 | Management time-line: software | 120 |
| 6.2 | SpiNNaker spike packet format | 123 |
| 6.3 | Schematic of the visualiser operation | 125 |
| 6.4 | Visualiser's right-mouse button menu system | 129 |
| 6.5 | Creation of neural networks for execution | 130 |
| 6.6 | Correct neural dynamics plot | 132 |
| 6.7 | Visualisation of erroneous single-unit neuron dynamics. | 133 |
| 6.8 | Synfire chain raster plot | 133 |
| 6.9 | Synfire chain population firing-rates | 134 |
| 6.10 | Interactive real-time neural network plots | 135 |
| 6.11 | V1 simulation screen-captures | 136 |

| | | |
|---|---|---|
| 6.12 | Oscillator on the NEF | 137 |
| 6.13 | Heat-map application visualisation | 139 |
| 6.14 | Heat-map channel temperatures over time | 140 |
| 6.15 | SpiNNaker core utilisation plots | 141 |
| 7.1 | Typical management software view | 151 |
| 7.2 | IP packet encapsulation of SNMP (version 2c) | 154 |
| 7.3 | Packet capture of SNMP get-response | 155 |
| 7.4 | MIB tree structure | 156 |
| 8.1 | Management time-line: system management | 161 |
| 8.2 | Memory map of a SpiNNaker MPSoC | 162 |
| 8.3 | The SpiNNmate translation service | 163 |
| 8.4 | Primitives within SDP packets | 164 |
| 8.5 | SpiNNaker MIB extract | 167 |
| 8.6 | MemGUI memory operations | 168 |
| 8.7 | SpiNNaker memory performance | 170 |
| 8.8 | DMA parameter impact on performance | 171 |
| 8.9 | DMA write performance based on different core / memory clocks | 172 |
| 8.10 | DMA read performance based on different core / memory clocks | 173 |
| 8.11 | System NoC arbitration tree | 174 |
| 8.12 | Aggregate System NoC fabric performance | 175 |
| 8.13 | SpiNNmate performance comparison | 177 |
| 8.14 | SNMP retrieved temperature data | 179 |
| 8.15 | SpiNNaker core loadings | 180 |
| 8.16 | Synfire chain spike-rates | 181 |
| 8.17 | Cacti plot of lab temperature | 182 |
| 8.18 | Alerting with Nagios | 183 |
| 9.1 | Completed Management time-line | 186 |

Tomographic slices through author's brain (flick-book)  odd pages

# Abstract

MANAGING A REAL-TIME MASSIVELY-PARALLEL
NEURAL ARCHITECTURE
James Cameron Patterson
A thesis submitted to the University of Manchester
for the degree of Doctor of Philosophy, 2012

A human brain has billions of processing elements operating simultaneously; the only practical way to model this computationally is with a massively-parallel computer. A computer on such a significant scale requires hundreds of thousands of interconnected processing elements, a complex environment which requires many levels of monitoring, management and control. Management begins from the moment power is applied and continues whilst the application software loads, executes, and the results are downloaded.

This is the story of the research and development of a framework of scalable management tools that support SpiNNaker, a novel computing architecture designed to model spiking neural networks of biologically-significant sizes.

This management framework provides solutions from the most fundamental set of power-on self-tests, through to complex, real-time monitoring of the health of the hardware and the software during simulation. The framework devised uses standard tools where appropriate, covering hardware up / down events and capacity information, through to bespoke software developed to provide real-time insight to neural network software operation across multiple levels of abstraction. With this layered management approach, users (or automated agents) have access to results dynamically and are able to make informed decisions on required actions in real-time.

# Declaration

No portion of the work referred to in this thesis has been submitted in support of an application for another degree or qualification of this or any other university or other institute of learning.

# Copyright

i. The author of this thesis (including any appendices and/or schedules to this thesis) owns certain copyright or related rights in it (the "Copyright") and s/he has given The University of Manchester certain rights to use such Copyright, including for administrative purposes.

ii. Copies of this thesis, either in full or in extracts and whether in hard or electronic copy, may be made **only** in accordance with the Copyright, Designs and Patents Act 1988 (as amended) and regulations issued under it or, where appropriate, in accordance with licensing agreements which the University has from time to time. This page must form part of any such copies made.

iii. The ownership of certain Copyright, patents, designs, trade marks and other intellectual property (the "Intellectual Property") and any reproductions of copyright works in the thesis, for example graphs and tables ("Reproductions"), which may be described in this thesis, may not be owned by the author and may be owned by third parties. Such Intellectual Property and Reproductions cannot and must not be made available for use without the prior written permission of the owner(s) of the relevant Intellectual Property and/or Reproductions.

iv. Further information on the conditions under which disclosure, publication and commercialisation of this thesis, the Copyright and any Intellectual Property and/or Reproductions described in it may take place is available in the University IP Policy (see http://documents.manchester.ac.uk/DocuInfo.aspx?DocID=487), in any relevant Thesis restriction declarations deposited in the University Library, The University Library's regulations (see http://www.manchester.ac.uk/library/aboutus/regulations) and in The University's policy on presentation of Theses.

# Acknowledgements

I wish to acknowledge all those who have supported me during my research years at the University of Manchester, and my decision to undertake a Ph.D.

Within the University, I'd like to thank all those around me, past and present, within the SpiNNaker and broader APT groups. They have provided great support, including answering my daft (and sometimes not-so-daft) questions with enthusiasm and patience. The team have been a great source of inspiration and challenges in all respects, and some have become badminton / drinking buddies. I'm particularly lucky to have chosen this group to join, it's an honour to have played a small part in the team developing the SpiNNaker architecture.

More specifically I'd like to particularly thank my supervisor Prof. Steve Furber, who responded positively to my initial contact in 2008, encouraged my application, and provided support all the way through the programme. Special thanks also go to Dr. J. V. Woods and Dr. J. D. Garside for their proof-reading and assistance in guiding the shape of this thesis into an examinable form.

In addition I greatly appreciate the support of the U.K. Medical Research Council (MRC), who provided the doctoral training award funding that kept me financially afloat over the Ph.D. programme.

# The Author

James *Cameron* Patterson graduated in 1996 from the University of Lancaster with a Bachelor of Engineering honours degree in Computer Systems Engineering.

He entered industry in 1996, in the telecommunications sector, predominantly working in public / private packet switched data networking environments specialising in Cisco Systems hardware platforms. Cameron worked for IBM's global network division from 1996-1999, and subsequently AT&T between 1999 and 2004 on numerous high-profile customer networks, and in a service and product development rôle for AT&T labs.

Professionally Cameron attained Chartered Engineer status in 2003 with the IET, and holds a variety of Cisco Systems certifications in design, security, voice and their practically examined high-level Cisco Certified Internetworking Engineer (CCIE) certification in routing and switching [Cis12b].

In 2004 Cameron needed a new challenge, so set up his own consultancy firm and contracted within the Edinburgh financial industry at Standard Life and the Royal Bank of Scotland in 'network convergence' support and advisory positions.

By 2008 Cameron had saved enough pennies to fund a number of months travelling around the world, and upon his return was successful in applying for, and beginning his Ph.D. programme in autumn 2008 at the University of Manchester.

# Chapter 1

# Introduction

In the race for improved computing performance, both overall computing capacity and energy efficiency play key rôles in the design and delivery of new computing systems. In very large-scale machines the processing hardware may be distributed across multiple cores, chips, boards, equipment racks, rooms and perhaps physical locations. Parallelism, once the preserve of research computing, is becoming the prevalent processing environment.

Parallelism gives rise to systems comprising large numbers of components, including processors, chips, power supplies, memory, disks, circuit boards and interfaces. The larger the number of elements forming a system, the greater the statistical likelihood of component failures at any time. It is therefore particularly important to manage the status of equipment and resources in such complicated machines.

SpiNNaker is a specialised massively-parallel computing architecture developed primarily to model large scale spiking neural networks in real-time. Its architecture is inspired by that of the biological brain, offering high-levels of computational parallelism, connectivity and redundancy in an energy efficient environment. This thesis covers the system management of SpiNNaker, within its complex (and constrained) real-time computing environment, which ultimately extends to million+ core machines. The management remit includes each processor's boot-up and testing, the loading and execution of software on the platform, through to the operational environment of the whole machine as it performs simulations – providing facilities for both hardware operators and software users to view the system and its performance in-flight.

## 1.1 Research Motivation

As the number of components rise in a computer platform, it may become difficult to attain a complete view of the operational status and serviceability of all parts of the system for the purposes of management [SyMAF00]. This is true for both hardware and software systems as they scale in size, including large deployments of the SpiNNaker platform.

### 1.1.1 Not Chips, but *Cores* with Everything

From the beginning of the integrated circuit era until the middle of the first decade of the 21st century, microprocessor architectures have typically used a single processor [Gee05]. Incremental performance gains centred around optimisation of this single processor – by reducing instruction cycles, adding pipeline stages, caching, increasing memory and miniaturising process geometry to permit increased clock speeds. As the majority of feasible performance gain using these techniques has already been attained – due to limiting factors such as power-consumption, heat-dissipation and leakage current – designers are increasingly turning to providing greater computational power by parallelism and placing multiple processors onto the same die [ONH+96]. Using this approach Chip Multi-Processors (CMPs) provide multiple execution streams and enhanced performance where the software being run is optimised / compiled to take advantage of multiple execution threads [Gee05]. However, not every application / system lends itself neatly to parallelism and multi-threading, and in a parallel environment this may leave computing resource sub-optimally used. The ever-increasing spiral of demand for more processing power is fast becoming more of a software than a hardware issue; to keep processors occupied, and managing limiting parallel factors such as inter-processor communication, memory input / output contention and coherence.

In CMPs there remains a physical space constraint on the chip die, so placing multiple cores in a single package is also limited in its scalability. Logically therefore, once core-numbers approach the limits on-die, multi-chip architectures are used to scale computer processing capacities further. Multi-chip solutions have been used in scientific high-performance computing for some time, with the same multi-threaded limiting factors, where effective performance is maximised by study and optimisation of the problem to limit the restricting cases. It remains to be seen whether sufficient parallelism can be incorporated into generic 'consumer' applications to make good use of many-core architectures.

Mainstream desktop and server architecture shipments now predominantly deliver CMPs, and this CMP approach even extends to the latest generation of mobile phone handsets and tablet computers [NVI12]. Amdahl's Law [Amd67] characterises the issue – illustrating that in multi-core systems the maximum speed-up achievable by parallelism is limited by the operational portion which may *not* be parallelised.

The 'Holy Grail' of efficient generic software parallelism seems no closer now than when the issues first arose and, whilst in the future large arrays of processors may be available, many of them may be idle at any one time unless there is a transformation in applications. Processors are becoming plentiful and a true commodity, and processing capability is increasingly not the limiting case in many environments, Amdahl's law is. Examining today's high-end mobile systems provides a valuable insight – aggressive power management turns off unused capacity to preserve the new measure of ubiquitous computing: *energy*.

**Approaches to Multiprocessor Computers**

Differing design philosophies may be taken when building high-performance multi-processor computers:

1. Computational Focus: use as many fast processors as required in an interconnected configuration to reach the performance target.
2. Energy Focus: deploy numerous, slower, energy-efficient processors to achieve the maximum performance possible, within the energy budget.

An example of the first approach, is the Blue Gene/P series of super-computers from IBM. The highest performing Blue Gene/P machine, JUGENE [JUL12], is recorded in the June 2012 TOP500 [MSDS12] (a list of the world's 500 most powerful computer systems). It is listed as attaining a peak of 825 TFLOPS (Linpack benchmark [DLP03]), using 2.268 MW (~364 MFLOPS / W) across nearly 300,000 PowerPC processors. As it becomes possible to build these increasingly large machines, their energy requirements begin to become a significant issue, logistically and financially. Wholesale electricity pricing in late 2010 (in the UK) was ~£50 per MWh [UK10]. The *wholesale* energy to run JUGENE therefore costs around £3,000 per day, £80,000 per month or £1,000,000 per year. The actual utility charges are likely to be far higher once the energy supplier margins and infrastructure costs are taken into account and added to the bill.

## 1.1. RESEARCH MOTIVATION

The next generation of the IBM Blue Gene family, the Blue Gene/Q, concentrates on greater energy efficiency. Machines using this architecture also have entries in the June 2012 TOP500 [MSDS12], but rather than JUGENE's 364 MFLOPS / W, the IBM Blue Gene/Q JUQUEEN (as an example, also installed in the Jülich Supercomputing Centre), attains 1,380 TFLOPS at ~2,000 MFLOPS / W. The Blue Gene/Q architecture is a chart topping entry in the June 2012 Green500 [FC12] (a list of the most efficient supercomputing platforms across the world), and is a significant generational advantage over its predecessor; being able replicate the same performance as the JUGENE system for around a fifth of the energy costs. The next major milestone in supercomputing is Exascale computing [Kot07], but even with the Blue Gene/Q generation of hardware Exascale computing is clearly infeasible, as the operational energy costs would exceed £200 million per annum, and require 500 MW to operate (a significant proportion of a large modern power plant).

SpiNNaker [PFT$^+$07] takes the second approach of using many less capable but highly-efficient processors, with core technology from ARM Ltd. operating at modest clock frequencies. This biologically inspired architecture can scale to a maximum of 65,536 chips, each with 18 ARM processing cores ($>$ 1 million cores in total), using 30-75 kW to deliver $>$250 TIPS with an efficiency bettering 3500 MIPS / W [PGP$^+$12].

The machines mentioned above are homogeneous devices, however there are alternatives to this approach using hybrid processing technology, such as the Kraken XT5, the Cray XT5-HE, the IBM Roadrunner, and Tianhe-1A [MSDS12]. A current, notable European initiative taking this hybrid approach is the Mont-Blanc project [dS12], initiated in October 2011 and coordinated by the Barcelona Supercomputing Centre. This 3-year European project aims to deliver the highest computing:energy ratio of any supercomputing platform. It, too, aims to take advantage of the ARM processor family, albeit in conjunction with nVidia GPU technology, and to attach boards modularly into larger systems. Their two prototype systems target 300 MFLOPS / W in the ARM only configuration, and with the ARM plus GPU version: 7,500 MFLOPS / W [Ini12].

Further approaches to multi-processor computing include building 'virtual' supercomputers using 'grid' technology [Fos03], incorporating a plethora of machines over multiple areas of control, or using clustering techniques such as Beowulf [SBS$^+$95].

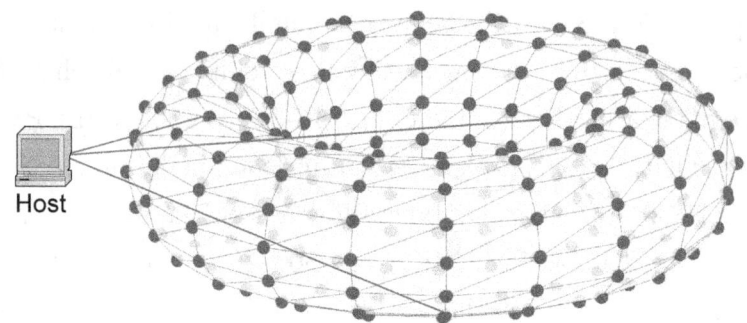

Figure 1.1: Multi-chip SpiNNaker CMP system shown as a torus.

### 1.1.2 Large System Management

With large parallel systems becoming increasingly prevalent, they require managing through all stages of their operation to validate the viability and correctness of the results they produce. Management of such a system begins at power up, whilst the hardware undergoes Power-On Self-Testing (POST) and system preparation, and continues during the loading of the system and throughout execution on the platform. Following completion of the job, the system continues to provide managed facilities including dumping data for further (off-line) analysis. The system is then either powered down or reset / reloaded and the cycle recurs.

Options for hardware monitoring of a large system are constrained [BAFL+11], and for conventional high-performance systems are predominantly based around bespoke tools supplied by the manufacturer, custom scripting developed by the system operators locally, and the use of generic agents and management software such as Nagios [BAFL+11]. Similarly, with software monitoring, the user is able to utilise the interface provided by the vendor, or may be able to modify existing tools or provide a custom solution.

### 1.1.3 Neural Networks on Parallel Computers

With its billions of processing elements operating simultaneously, simulation of the brain seems an ideal fit for the parallel computation model. The brain consists of highly interconnected repeated fundamental computation units (neurons), which act together to form an energy-efficient, high-performance computational platform. In an Artificial Neural Network (ANN) each of the simulated neurons and associated inputs may be modelled on available processors, including the ability to statistically multiplex neurons onto the computing resources as the neural model runs in parallel. The

## 1.1. RESEARCH MOTIVATION

number of neurons that may be modelled depends on the biological fidelity required of the simulation, the target operational time-base and the parallel computational resource available. As each biological neuron operates asynchronously to the system as a whole, without reference to a coherent clock or memory store, ANNs offer few of the serial bottlenecks of conventional parallel programming and high resource utilisation can be attained. Artificial Neural Network modelling does, however, rely on rich networking to replicate the high-degree of interconnection found between biological neurons. The capacity of the communications fabric (measured as its bisectional bandwidth) must be sufficient to convey the large number of neural signals (spikes) as they are emitted and replicated in the network, fanning-out to all destination neurons. This characteristic (and particularly the small message types used to represent spikes) is often problematic for conventional parallel platforms, leading to the development of specialised architectures.

### 1.1.4 Monitoring a Specialised Massively-Parallel Neural System – SpiNNaker

SpiNNaker is an novel, scalable architecture designed to extend up to a million processors by the interconnection of SpiNNaker chips. There are 18 low-power ARM cores in each SpiNNaker chip, the chips typically interconnected in a torus configuration (fig.1.1). This configuration of a SpiNNaker machine provides a massively-parallel, resilient environment, ideally suited to the simulation of large artificial neural networks.

Whilst the new bespoke SpiNNaker platform is in its infancy there is no pre-existing software in its management realm. The SpiNNaker platform is a resource-constrained one, with modest amounts of memory and processors more frequently found in embedded environments than in high-performance computers; this provides a limited set of paths which may be taken for the management platform. 'Customers' of the SpiNNaker machine will naturally assume that manageability features will be provided, and this requirement drives much of the research detailed in this thesis. Its goal is to research and develop frameworks for scalable systems management of the SpiNNaker machine, throughout its full run-time cycle.

Figure 1.2 depicts the areas over which the management requirements of SpiNNaker overarch, which touch on both its hardware and software. This thesis follows a chronological path in the system management time-line of a SpiNNaker simulation, and figure 1.2 and derivatives thereof are used throughout to indicate in which area each chapter's contributions are made.

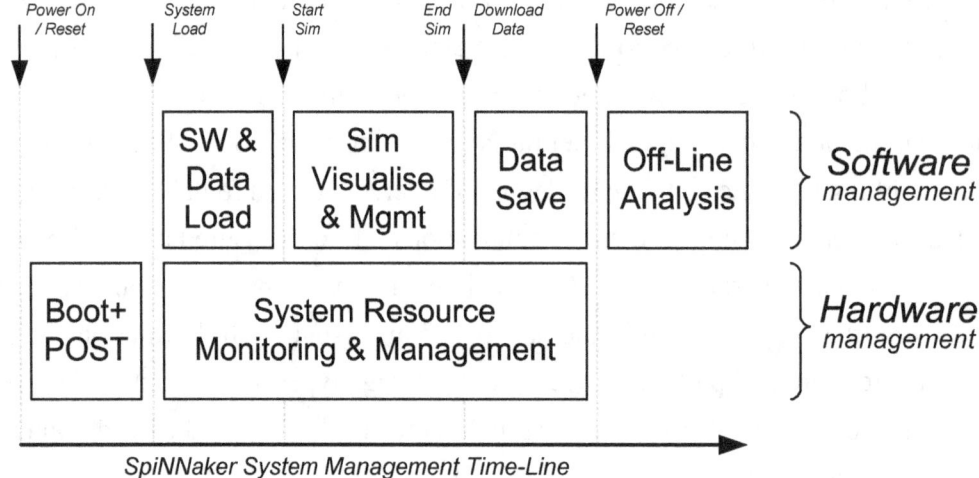

Figure 1.2: The system management time-line of SpiNNaker.

The SpiNNaker management solution is required to scale to a million processor machine and operate in real-time, from power-on through reset cycles to final power off. It is required to provide facilities for hardware operators so that faults within the system can be identified, mapped out if possible, and rectified in a timely fashion. SpiNNaker has a much wider set of users than just hardware operators, as it is not designed to operate as a 'black-box' batch execution system. Researchers running software on the platform will likely come from areas including psychology, computational neuroscience and machine learning – therefore providing these user groups with a suitable management path is also necessary. As SpiNNaker is built with the main purpose of operating massive artificial spiking neural networks in real-time for the user groups indicated, a software visualisation interface which provides emulation of existing biological functional brain imaging techniques such as fMRI, PET and EEG may fulfil their requirements. However as ANN simulations operate on a computational platform, it is possible to retrieve fine-grained noise-free detail through simulation imaging, compared with the biology, helping to gain greater functional insight. These detailed visualisations can be gathered at the level of single neurons and synapses, through assemblies and populations, to entire aggregated network representations.

A large SpiNNaker deployment retains an immense amount of state. If this data were to be downloaded in a single session, then the output paths would be saturated by a million+ cores, 200,000 inter-chip links, and approaching 10 terabytes of RAM. For this reason the management system must focus its monitoring – initiating aggregation and sampling techniques to ensure it does not impact the simulation being run. When looking at diagnostic information it should only enable functionality in areas where required: for ongoing statistics, or when a problem is detected in an area and the management agent is 'zooming' in to see the detailed diagnostics. The user viewpoint needs to be able to encompass the full range of granularity to look at the system health / performance as a whole, by neuron population, per chip, core or neuron level. Given that the simulation is designed to emulate up to 1 billion neurons in real-time, this range needs to support scales of $>10^9$:1.

While it is clear that the external links are constrained by the channel bandwidth available to them, there are also significant limitations on the processing cycles available to provide management functionality. Each SpiNNaker core has small local memories, but has the primary task of simulating neural networks, therefore the lightest of touches is a fundamental requirement of the management system applied to the machine. This applies in terms of memory footprint, processor cycles and bandwidth use, to ensure the primary application function is not disrupted.

Whilst the motivational focus of this thesis is the SpiNNaker system, a lightweight management framework (developed in this thesis) may also be applicable to other large resource-constrained distributed applications. This is particularly prescient as ARM core technology (as used by SpiNNaker) is widespread, with over 20 billion licensed ARM cores deployed across the globe by 2012 [ARM12]. ARM technology at present finds itself primarily in (constrained) mobile and embedded environments, a similar situation to which each SpiNNaker processor finds itself within.

## 1.2 Contributions

There are a number of contributions the author has made forming the SpiNNaker management framework time-line (fig. 1.2) and chronologically these include:

- Within the SpiNNaker chip silicon there resides a ROM which has the software used to initialise and test each chip, processor and peripheral. This ROM image contains novel routines to reliably flood-fill software throughout the machine, for dynamic fault recovery, and to deal with network communications across a machine that powers up with tens of thousands of homogeneous hardware components. Results from this work have been published as [SPF11].
- Within the area of software monitoring a primary data visualisation tool has been developed for managing simulations executing on the SpiNNaker machine. The

visualiser works in real-time, plotting activity and providing facilities to interact with the neural network simulations. Visualisation techniques have been tailored to the users' requirements, and to suit the SpiNNaker execution environment, including the ability to extend to new ANNs (and to plot non-neural data). The results from this contribution have been presented at the International Joint Conference on Neural Networks 2012 [PGRF12], and are covered in more detail in chapter 6.

- Within the SpiNNaker system hardware monitoring, the main contribution is the protocol conversion to permit standard tools to be used to manage the SpiNNaker platform. This concept enables tools to interface with the SpiNNaker system through a single, unified, and extensible, communications framework. This novel translation function additionally links with databases to perform mapping between hardware resources and software functions using them. This technique also permits extension of the monitoring technique from hardware, to software components in the system. The results of this contribution are covered in more detail in chapter 8 and have been accepted for the 2012 EuroMicro Digital Systems Design conference [PPGF12].

The SpiNNaker project is very much a team effort, and while the above areas within the management realm are the largest contributions led by the author, other contributions have been made with simulation, testing and specifications of the SpiNNaker chip, and in areas and projects led by other group members. Where this work led to specific publications, this is indicated in the publications section below. Similarly within the management areas which I have led, there have been contributions made by other team members acting from the author's lead. These contributions where substantive are indicated.

## 1.3 Publications

The following is a chronological list of major peer-reviewed publications to which the author has contributed. Each publication is referenced, and the contribution to each is specified:

- ACM Computing Frontiers 2010 conference paper, "Scalable Event-Driven Native Parallel Processing: The SpiNNaker Neuromimetic System" [RJG$^+$10]. Contribution: Real-Time Communications between the Host System and the

SpiNNaker board to enable to control of the Hunter in the simulation environment.

- IEEE World Congress on Computational Intelligence (WCCI) 2010 – International Joint Conference on Neural Networks (IJCNN) conference paper, "Algorithm and Software for Simulation of Spiking Neural Networks on the Multi-Chip SpiNNaker System" [JGP+10].
  Contribution: Support of real-time data collection, by enabling spikes to be sent to / aggregated and received from the SpiNNaker system.

- International Conference on Neural Information Processing (ICONIP) 2010 conference paper, and Australian Journal of Intelligent Information Processing Systems, "Interfacing Real-Time Spiking I / O with the SpiNNaker Neuromimetic Architecture" [DPG+10].
  Contribution: Software and control algorithms, communication and interfacing of silicon retina and robot to the SpiNNaker system.

- WCCI 2011– IJCNN conference paper, "Distributed Configuration of Massively-Parallel Simulation on SpiNNaker Neuromorphic Hardware" [SPF11].
  Contribution: Led on the boot software, flood-fill algorithms and error checking.

- Parallel Computing article, "Event-Driven Configuration of a Neural Network CMP System over an Homogeneous Interconnect Fabric" [KRN+11].
  Contribution: Boot up protocol and application loading process.

- International Journal of Parallel Programming article, "Managing Burstiness and Scalability in Event-Driven Models on the SpiNNaker Neuromimetic System" [RNJ+11].
  Contribution: Spiking model communications I / O setup in hardware and in silicon simulation.

- In press – Journal of Parallel and Distributed Computing article, "Scalable Communications for a Million-Core Neural Processing Architecture" [PGP+12].
  Contribution: Lead author. This article contains material based on a subset of chapter 4 and is copyright Elsevier.

- WCCI 2012 – IJCNN conference paper, "Visualising Large-Scale Neural Network Models in Real-Time" [PGRF12].
  Contribution: Lead author. This paper contains extracts from chapter 6 and is copyright IEEE.

- Accepted for 15th EUROMICRO 2012 Conference on Digital System Design, "Managing a Massively-Parallel Resource-Constrained Computing Architecture"

[PPGF12].

Contribution: Lead author. This paper is based on some of the material forming chapter 8, and is copyright IEEE.

- Accepted for 2012 IEEE Custom Integrated Circuits Conference, "SpiNNaker: A Multi-Core System-on-Chip for Massively-Parallel Neural Net Simulation" [PPG$^+$12].

Contribution: Memory performance results as covered in chapter 8 of this thesis.

## 1.4 Thesis Overview

Following this introductory chapter the thesis continues with a grounding on the essential biology of neural networks and background on the development of artificial neural networks, and the systems on which they run. The thesis structure then follows the chronology of the SpiNNaker system management time-line (fig. 1.2). The time-line items are covered in turn, with a preceding review chapter followed by methods, tools and results. This chronological order provides a logical flow from beginning to end of the time-line of the SpiNNaker management framework. The thesis closes with a discussion chapter which draws conclusions from the SpiNNaker management research. A brief overview of each individual chapter follows:

### Chapter 2 – Neural Computing

This chapter covers the basic cell and connectivity biology of the brain which is repeated billions of times in the brain to form neural networks. The chapter then explores the development of the different generations of *artificial* neural networks, from the simple threshold logic units, through to the next generation of multi-layered perceptron models incorporating learning models, and finally to the third generation: Spiking Neural Networks. The chapter concludes by exploring hardware that has been used to undertake artificial neural network simulations, particularly focusing on the larger scale undertakings, from the general purpose to the bespoke.

### Chapter 3 – The SpiNNaker Neural Architecture

Chapter three describes SpiNNaker, a large-scale artificial neural network simulation architecture being developed by the Universities of Manchester and Southampton, and

## 1.4. THESIS OVERVIEW

partners. This chapter explores SpiNNaker's features and specifications, particularly as they relate to system manageability, the available interconnectivity and software.

### Chapter 4 – Bootstrapping SpiNNaker

Chapter four covers the creation, implementation and results of the Node-Boot ROM software executed at power-up or reset of each SpiNNaker chip. This includes the management aspects of the design process from the power-on self-tests and fault isolation, to the loading of the application software. Detection of capabilities and system health of the components at this stage aids in the mapping and assignment of load to the system. Testing of the ROM is carried out and its performance is evaluated based on its initial requirements. The author led the research and development in this area, and contributed around two thirds of the programming effort and testing, with the balance performed by Thomas Sharp from the SpiNNaker research group.

### Chapter 5 – Imaging Neural Networks

This chapter firstly reviews the techniques available to record and image the human brain, covering both its anatomy and its activity, and combinations thereof. The chapter then moves on to cover visualisation techniques used within the field of artificial neural networks, and how they differ between generational types. Tool-sets and analysis techniques for both 'wetware' and artificial neural networks are reviewed to determine the most popular and useful visualisation techniques.

### Chapter 6 – Visualising Neural Networks on SpiNNaker

Chapter six covers the specification, implementation and results of a novel real-time visualisation tool tailored to the requirements of the SpiNNaker system and users. Visualisation techniques identified in the previous chapter are created, together with an interface for the user to be able to interact with their simulation. Scalability is a key topic within very large neural networks, and techniques are devised for data aggregation and for turning data collection on and off appropriately. As a single parameter on a particular emulated neuron may be as important as the high-level overview of the system, techniques covering the whole range of scales in navigation are explored.

## Chapter 7 – Managing Large Network Attached Systems

In this chapter management techniques for large and network attached computing systems are reviewed. The chapter explores monitoring of traditional and bespoke computing platforms, and the FCAPS systems management framework devised by the ISO. A review of common management data structures and protocols is carried out and a more detailed analysis made of the successful SNMP standard and how it is applied to network attached systems.

## Chapter 8 – SpiNNaker Management Framework

In chapter eight the philosophy of efficiently monitoring the large neural computer, SpiNNaker, is explored, particularly given its resource-constrained execution environment. It is argued that different sets of requirements can actually be catered for using the same simple underlying network management framework. This framework provides an abstraction of the detail from the machine hardware to objects via a translation function. The translation function is implemented and tested in conjunction with the industry standard management protocol SNMP, to produce a system that allows SpiNNaker to be monitored by commercial off-the-shelf software, but is provided at a low implementation cost to the SpiNNaker machine itself. The author led the research in this area, and was assisted by a third year undergraduate student, Thomas Preston, who carried out much of the host-side programming work.

## Chapter 9 – Discussion and Conclusions

In this chapter each of the contributions to the management time-line as implemented on SpiNNaker is discussed and evaluated. As the machines grow in size the scalability of the solutions is examined, and future enhancements of each of the components are suggested. This chapter contains overall conclusions for the SpiNNaker management framework – it explains how standard tools can be applied to the management of a bespoke machine efficiently, how ANNs can be visualised and interacted with in real-time on the system, and how the boot process is successful in providing a managed platform on which the rest of the system may stand.

# Chapter 2

# Neural Computing

Scientific study of the brain has been undertaken for well over a hundred years in the attempt to comprehend its function and form, and to aid in the treatment of its ailments. The brain is a massively-parallel processing organ and offers large resilient computational ability, albeit provided by a hugely complex system in a far from transparent fashion. Biological brains exceed any electronic computational platform in energy efficiency and parallelism and exhibit the remarkable ability to continue to operate effectively even whilst impaired.

Conventional neuroscience still has a key obstacle in understanding the brain – its work is typically invasive and destructive. Much research into the brain is performed in-vivo or post-mortem, with inevitable drawbacks. With post-mortem study, cellular-decomposition and the lack of natural stimulus from the host are clearly limiting; in-vivo has ethical considerations, together with an inability to reach every part of the neural system for study, at least without causing damage to the organ as a whole. In-vitro experimentation is an alternative, however experimentation is limited to cells and clusters which may be cultured, and again lacks the real animal input and stimulus – there is no 'closed loop'.

Each new generation of computational hardware, however, advances the possibility of exploiting increased processing and communications capacity for simulating brain function and pathology by computer experimentation. The biological function of the brain itself is also of considerable interest to computer and software designers who are coming to terms with the explosion of parallelism, now required, to continue the exponential performance gains expected of them.

This chapter continues with some biological concepts of neural networks which are applicable across many chapters of this thesis, and explores what is known about the

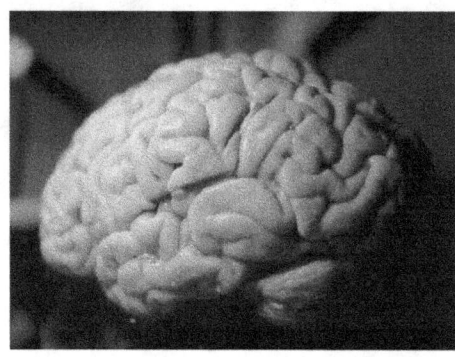 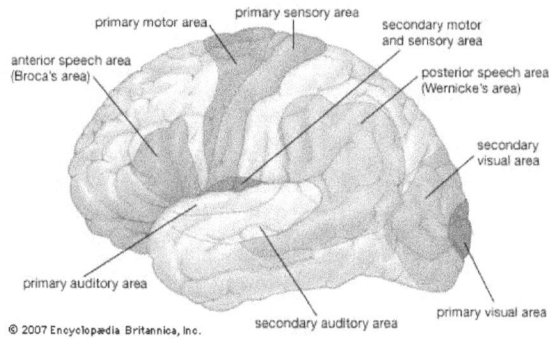

(a) Human brain (src: Wellcome Images)  (b) Common cortex regions (src: [Onl12])

Figure 2.1: The human brain.

brain today as it applies to simulations. The latter part of the chapter covers attempts to simulate fundamental brain components and functions in electronic form, covering both the software models and the hardware on which they run.

## 2.1 Biology

The mammalian brain is one of the most complex structures known in the universe, and there is still no real consensus on how it works as a whole [Uni12]. Experimentation in the 19th and early 20th centuries has already provided insight into the physical make-up of the brain, and by observation of dissected samples across many species the fundamental components have been identified. The outer layer of the brain comprises the cerebral cortex, which in more complex organisms is folded into the familiar form found in figure 2.1a, providing an overall larger cortical area. In humans, the typical cortical surface area is around 0.25 m$^2$ [JL07]. Functional areas have been broadly identified and mapped onto a cortical patchwork on the brain (Brodmann [Bro09]), a simplified annotated example can be found in figure 2.1b. Different species, including humans, have distinct functional regions corresponding to their needs. The human cortex primarily follows a regular 6-layered neuronal structure [Mou78, DM91] forming a 'cortical column' of around 3 mm in depth (fig. 2.2)*. Regular structures can be identified amongst the cortex forming mini-columns of around 100 neurons apiece, approximately 100 of which comprise a hyper-column. Within this model, the human

---

*Note: The notion of the cortical column is contentious, with some researchers taking the position that there is no such regular cellular 'micro-circuitry' within the brain.

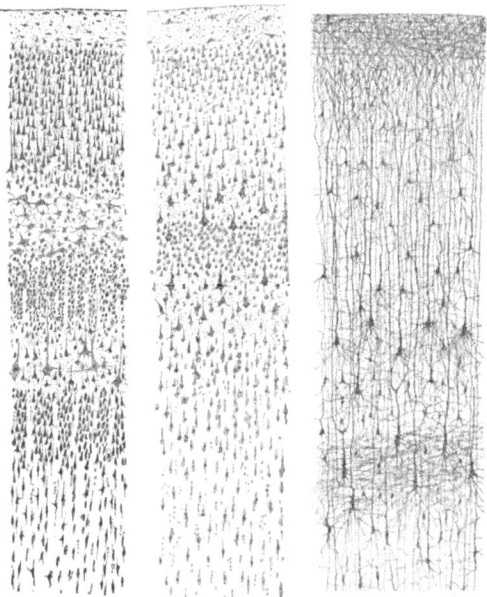

Figure 2.2: Cajal's classical drawings of the six-layer human cortical column (shown above the connective white matter at the base of the figure) (src: [Caj99]).

brain is thought to have around 2 million hyper-columns [JL07], and around 200 billion neurons. Other estimates of total neuron count do vary [WH88, HH09], but as the brain is so complex, each is a best estimate dependent on the methodology used.

## The Neuron

The fundamental computational unit in the brain is the *neuron* [Tho00], which is interconnected with many other neurons in the cerebal cortex typically forming cortical columns. A simplified structure of the neuron and its major components can be found in figure 2.3. Each neuron cell is formed around a nucleus which is enclosed within a cell body called the soma. Connections are made with upstream *afferent* neurons via the dendritic tree, and electrical inputs arrive from this tree at the neuron's cell body. These inputs, in conjunction with the current polarisation of the cell (the membrane potential), determine whether the neuron should emit a spike (also known as an action potential). The cell's spikes are transmitted along the neuron's axon output, which may extend some distance and is usually sheathed in a fatty insulating material known as myelin. Myelinated construction of the axon results in accelerated, but relatively slow propagation of the spike, ranging to over 100 metres per second [Wax80]. Branches are taken from the neuron's axon, and its terminals connect with the dendrites of other downstream *efferent* neurons, the junction of which is known as the *synapse* (fig. 2.3).

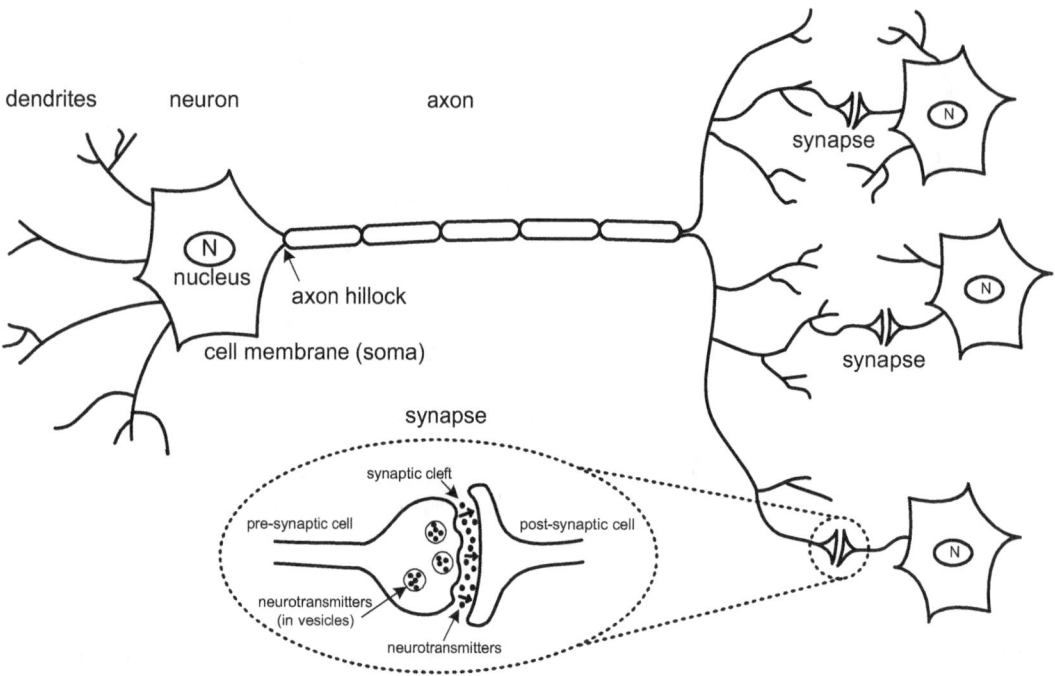

Figure 2.3: Simplified neuronal structure.

## The Synapse

A synapse is an electro-chemical junction where action potentials are decoupled between neurons across a synaptic cleft. In the predominating chemical synapses, depolarisation following receipt of an action potential leads to transmission of neurotransmitting chemicals across the synaptic cleft between the pre-synaptic axon terminal and post-synaptic receptor. This process can be either excitatory or inhibitory (increasing or decreasing) the post-synaptic neuron's membrane potential. Within the human brain it is estimated there are around 1,000 synapses for each neuron (as a gross average, some have *many* more and some *far* fewer), totalling $10^{14}$ synaptic connections [NLMA+09]. In the simple example found in figure 2.3, for clarity, there are far fewer connections shown than are typically found for an actual neuron where a connection fan would be in the hundreds. These numbers multiplied over a large neural network, such as a brain, would form an almost insurmountable connectivity problem if the connectivity were random. However it is typically found that neurons have a high density of local connections, with longer connections being somewhat more sparse.

### Neural Plasticity

Given that the sum of inputs and the current state of a neuron combine to determine whether that particular neuron should 'fire', there must be a mechanism which permits the influence of each input to be altered depending upon its (varying) importance. This mechanism is known as plasticity, and it permits connections between a pair of neurons to be strengthened or weakened depending on the correlation between one another's spiking behaviour. The changes in behaviour are triggered chemically in the synaptic connection between two neurons, and can be related to 'memory' as the weightings between the two neurons are persistent to a greater or lesser extent over time. For this reason the area of learning, particularly synaptic plasticity and how it changes over time, has received a great deal of study after a conjecture in the late 1940s by Hebb [Heb49].

In Bliss and Lomo's work [BL73], evidence for the chemical operation of Long Term Potentiation (LTP) was discovered in a study with rabbits. This mechanism, in conjunction with its converse, Long Term Depression (LTD) are proposed plasticity models which occur within synapses. The consequences are an adjusted behaviour affecting how incoming spikes are passed on to the post-synaptic neuron. Another biologically plausible model of plasticity is described by Spike Timing Dependent Plasticity (STDP) [AN00], which is found to match the behaviour of some types of synapses. STDP relies on the timing of ingress and egress spikes of a neuron to determine whether to strengthen or weaken a synaptic connection. If the pre-synaptic neuron has fired and this input arrives in advance of the post-synaptic neuron firing, the correlation is used to strengthen the connection as they appear to be causally linked. Similarly if the input spike arrives shortly subsequent to a neuron firing then it is less likely to have been related, so the strength of that link is depressed. The impact of these changes is additionally related to time, with closer spike time correlations affecting larger proportional changes in the weighting.

## 2.2 Neural Network Modelling

From an early stage it has been postulated that the fundamental elements of the brain can be mathematically modelled, to create *artificial neural networks* (ANNs). The apparent simplicity, ubiquity and regularity of the neuron model makes it highly desirable to create electronic simulations; to study their behaviour to devise improved treatment regimes medically and to understand more of the brain's computational ability.

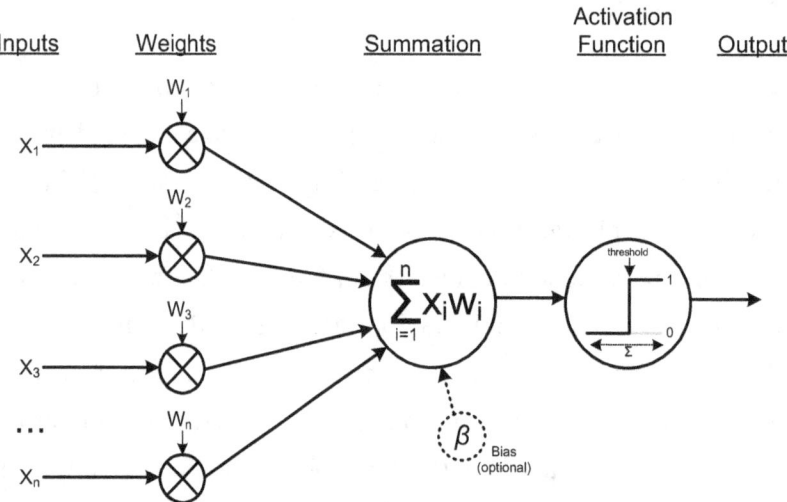

Figure 2.4: Threshold Logic Unit (TLU). When the sum of inputs crosses a threshold the TLU outputs a 1, otherwise the output is zero.

### 2.2.1 First Generation Artificial Neural Networks (ANNs)

The field of ANNs was spawned with the work of McCulloch and Pitts in 1943 [MP43] in proposing the Threshold Logic Unit (TLU) for artificial neurons. In a TLU neuron inputs are weighted and summed, and compared to a simple threshold value to determine whether to emit an output (fig. 2.4). Their work was based on simplified observations from 'wet' neuroscience, and gave a foundation to the implementation of artificial neural networks using digital electronic and computing platforms. The TLU technique, however, is an artificial approximation – it defines that when a neuron is activated it can emit only a Boolean 1 or 0 to its downstream connections – without regard to time, or the actual continuous dynamics of biological neurons. It should however be noted that this initial generation of ANN, albeit crudely defined, is still in widespread use, extended into networks such as perceptrons for pattern matching [Res58]. The sum may be optionally (and artificially) biased to influence the result of the sum if required, rather than adjusting the threshold, and this same external bias may be applied to many TLUs simultaneously.

### 2.2.2 Second Generation ANNs

In the first generation of networks there is a limitation based on the requirements for binary inputs and outputs of the system, and its static nature. In the second generation of artificial neural networks time now plays a key rôle, in conjunction with a non-binary

## 2.2. NEURAL NETWORK MODELLING

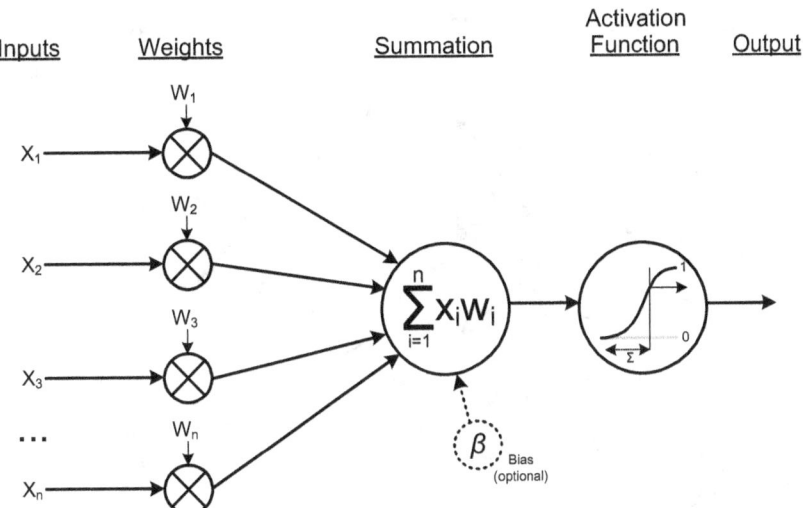

Figure 2.5: A second generation Artificial Neural Network (ANN), with a sigmoidal activation function. The output is continuous based on the inputs received.

activation function [Bis96]. In this model it is the *rate* of spikes output from a neuron which is believed to encode the information, and to represent this rate a continuous activation function (such as a sigmoid) is applied. Inputs to a neuron are again weighted and summed (with optional bias), but instead of a binary result from the comparison with a threshold, the output is determined by the position of the sum on the continuous sigmoidal output curve (fig. 2.5). In practice the shape of the sigmoid may be scaled as appropriate to make it shallower or steeper, and to speed up computation the curve function may be approximated linearly. The output of a second generation neuron can therefore be considered a probabilistic representation of the biological firing-rate of a neuron population, should it be required to recreate 'spike trains' from its output.

One of the greatest advantages of the second generation of neural networks is that they are capable of employing 'learning' techniques – using feedback to converge to the problem space. The primary technique used in this space is Back Propagation originally discovered in 1974 by Werbos, published in his Ph.D. dissertation [Wer74], but not exploited fully until the mid 1980s. In this technique cycles of operation are employed and weights in a Multi-Layer Perceptron (MLP) (fig. 2.6) are adjusted based on the error passed back from the downstream layer for each network iteration (or epoch). It was only around 1989 that Heicht-Neilsen [HN89, HN90] showed that a 3-layer MLP can solve non-linear separation problems, such as XOR. These results were able to truly refute the criticism of ANNs in the 1969 Minsky and Papert book [MP69], which had stifled research and funding for the area for many years.

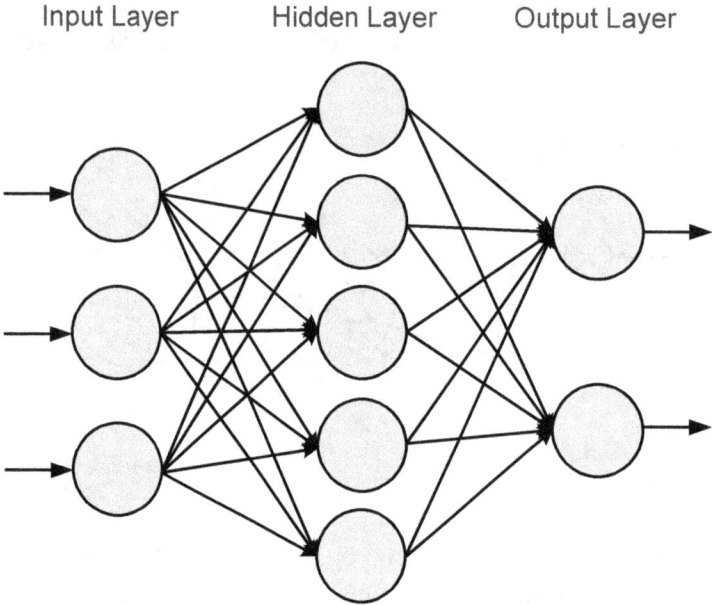

Figure 2.6: A Multi-Layer Perceptron (MLP) with 3 layers (input, hidden and output).

A further advantage of second generations networks is that they encompass the functionality of the Boolean first generation of ANNs [Maa96]. Second generation modelling remains commonplace [Bis96], particularly Multi-Layer Perceptrons (MLP) with back-propagation, both in software and distilled to hardware, but there are classes of problem that prove intractable using this technique [Maa96].

### 2.2.3 Third Generation ANNs

Latterly the *third generation* of artificial neural networks has become popular, based not on the output *rate* of the spike train, but on the temporal information (i.e. the *arrival times*) of the input spikes themselves. In this model the operation of the neurons is decoupled from a system-wide synchronous update cycle, which is far more biologically faithful to the spiking model as seen by Hodgkin – Huxley in 1952 [HH52]. Izhikevich [Izh04] illustrates a diverse range of neuronal spiking patterns replicating many of those found in biology (fig. 2.7), illustrating clearly why simple rate-coding may not be sufficient for all modelling purposes.

Indeed, rate-based techniques have been found inadequate to model visual analysis functions, for example, as they cannot achieve the response times demonstrated in the biological brain [Maa96]. By choosing to integrate the inputs of a neuron and analyse their rate-of-change, it is possible to examine the relationship of the temporal spacing

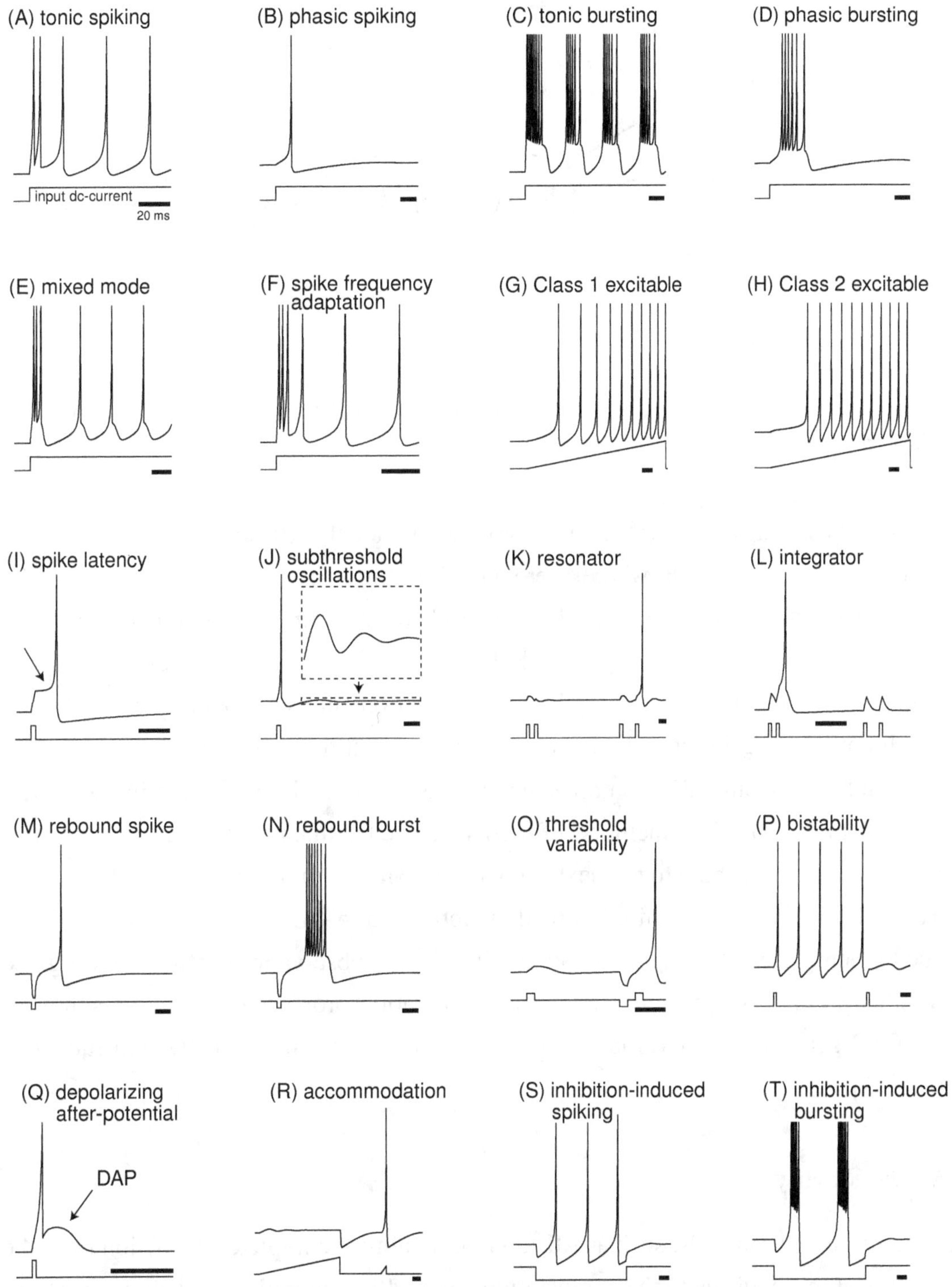

## 2.2. NEURAL NETWORK MODELLING

Figure 2.7: Spiking behaviours achievable when using the Izhikevich spiking neuron model [Izh04] (src: Electronic version of the figure and reproduction permissions are freely available at http://www.izhikevich.com).

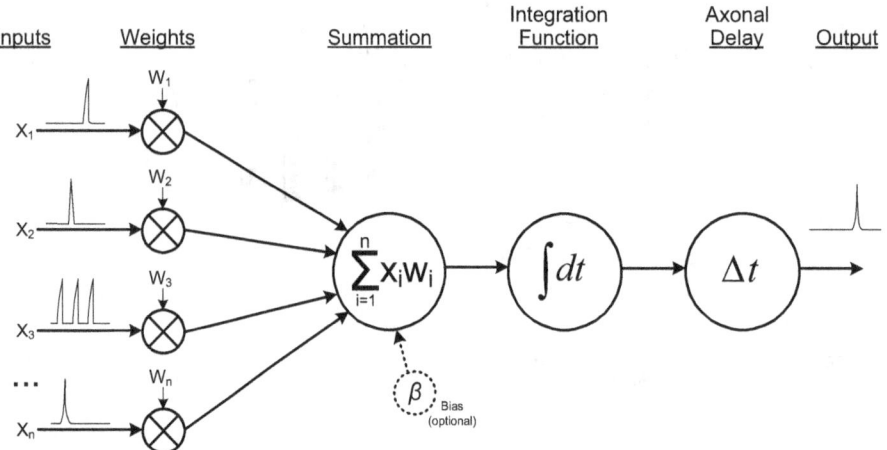

Figure 2.8: A third generation ANN, spike time arrival is now the key function (src: adapted from [FT06]).

of the spikes, and use this as the basis for network models. These techniques are called 'integrate and fire' models, where afferent neurons provide activation stimuli, and if the neuron's membrane potential threshold is reached (including any optional bias to this potential) a spike is emitted (fig. 2.8). Following a spike event the neuron's membrane potential resets to its resting potential, and the neuron enters a refractory period where it will not spike again regardless of how much stimulation it receives.

If during operation the membrane potential accumulated from input stimuli decays, providing a diminishing 'memory' of inputs received at that neuron over time, and this potential may contribute to the next firing event. Models employing this technique are called 'leaky integrate and fire', or by their abbreviation – LIF and are commonly used due to their good output approximation of biological observations. Maass investigated whether second (and by implication first) generation problem spaces are a subset of the third generation's, and found it to be the case for sigmoidal activation functions [Maa97].

**Model Fidelity**

In some network models, such as LIF, there is an almost complete abstraction from the biological mechanisms to the computational model. Adding biochemical fidelity typically involves reducing either the size of network that may be created, or the speed at which it may be simulated (or both), on a given simulation platform. Different projects and protagonists have differing priorities and approaches to the fidelity problem, trading off the different drawbacks and benefits.

In 1952 Hodgkin – Huxley [HH52] determined a detailed model of synaptic ionic dynamics which is often used to maximise biological fidelity, but this comes at the cost of large computational expense. Projects such as Blue Brain [Mar06], take the high-fidelity approach across all network components to achieve the closest possible approximation to biology. Others use highly simplified models including LIF variants such as the Izhikevich model [Izh03], believing that the dynamics of the neural system may be simplified into their most basic fundamental behaviours. There are of course a myriad of approaches that lie between each of these extremes, Izhikevich detailed many of these in his 2004 review paper 'Which model to use for cortical spiking neurons?' [Izh04].

**Synapses and Learning**

As with neurons, there are multiple synaptic models available for use in simulation [DA01]. Synapse models are an important part of neural network modelling, learning and plasticity, transforming spikes from pre-synaptic neurons into electrical stimuli delivered to the dendrites of downstream neurons. In earlier artificial neural network examples the synapses were approximated by weight functions (see figures 2.4, 2.5 and 2.8) with no mention of how the weights are calculated. Weights may be set statically, or learning may be used to trigger plasticity models such as LTP, LTD and STDP which have been adapted into ANN algorithms. To enable a network to function correctly, early adjustment of the synaptic weights (training) may be employed, where weights are adjusted interactively based on one of three learning categories: supervised, unsupervised and reinforcement. It is important not to 'over-train' a network, as it may learn to fit to the training data too well, and lose its capability to generalise.

## 2.3 Systems for Neural Network Modelling

Many different system architectures have, and are, used for neural modelling, including a range of dedicated, fixed, and programmable hardware, using both analogue and digital technologies – and those which use general-purpose hardware, from supercomputers, to clusters and individual workstations. While concentrating on the more scalable solutions, a number of high-profile projects are covered in detail by de Garis et al. with their world survey of artificial brain projects [dGSGR10].

Logically, applying custom hardware solutions to the problem of neuron modelling seems to offer the best solution in terms of power and resource utilisation. However, there exists a wide gulf in opinion as to which neural network models provide the appropriate level of biological fidelity [Izh04]. This opinion ranges from detailed models, which include full biochemical ion channel dynamics, to abstracted LIF neurons which are far less complex computationally. For this reason a wide range of hardware, software and hybrid solutions have been developed in the ANN space, and this section covers the background, advantages, disadvantages and scalability of the approaches.

Modelling neural networks is particularly hard on the communication infrastructures of machines performing large scale simulation. As a practical example of the volume of traffic large networks generate: consider a billion neurons, each with a biologically plausible average 10 Hz spiking rate and fan-out of 1,000 synapses. The model needs to support 10 trillion communication events per second (at whatever overhead the platform imposes for each spike). The pre-requisite for being able to cope with great quantities of communication traffic in neural network simulations has been outlined by Hines et al. [HKS11].

### 2.3.1 General Purpose Hardware

In this section the approaches taken to utilise general purpose computing hardware to execute ANN simulations are considered. As the possibilities for single processor improvement have been hampered by limiting factors such as feature geometry, heat dissipation and power provision, the rise in computational power is now primarily driven by parallel processing. This takes the form of multiple processors per chip, multiple chips – or both. Fortunately for artificial neural networks, the discrete elements used to construct ANNs fit neatly into the parallelisation of simulation. Using general purpose hardware has a clear advantage in that the computational power available in this space grows year-on-year driven by industrial development, whereas a dedicated hardware implementation has its potential fixed at the time it is committed to manufacture.

**Software**

There are many neural network modelling packages which operate across a wide spectrum of general purpose hardware platforms. A 2011 survey published in Frontiers in Neuroinformatics [HH11] suggests that the majority of users operate a GNU / Linux environment, and the authors provide a website of supplementary analysis [Neu11] detailing the software packages most popularly used for neural modelling:

- NEURON [HC97] is a simulation environment typically used to model single, or networks of, neurons. It abstracts the computational environment on which the simulations run from the specification and (if required) mathematical detail, with extensive support for graphical tools to be able to create networks using drag-and-drop techniques. NEURON can be run on a multitude of operating systems and environments, including clusters and high performance machines. As a gauge of the popularity of the tool, the NEURON team announced in January 2012 via their blog that 1,176 peer-reviewed publications include work performed on the NEURON platform. NEURON is especially strong in the areas of high biological fidelity, the wide-ranging availability of library components and where extensibility is required via techniques such as scripting.

- Brian [Goo08] is a spiking neural network simulator which may be used on a range of commonly deployed operating environments. It uses Python [Fou12] as a scripting language to create its simulation specifications and neural equations, enabling fast model development and testing within a single software tool. Whilst Brian is not as abstracted from the detailed implementation as NEURON at the outset, the familiarity of the Python environment is useful for those with computational backgrounds, providing simple and ready access to all aspects and state within the simulation. Brian supports the utilisation of parallel environments in a very limited way, in that separate instances of the simulation may be run, but not in concert with one another. This is a significant disadvantage as the majority of platforms now deployed use parallel processing via multiple cores. Users are advised to look elsewhere for large scale simulations or those with complex biophysical requirements.

- PyNN [DBE+09] is a Python based tool which is used for the description of neural network modes, but does not contain its own engine for simulation, instead using NEURON, NEST, PCSIM and Brian for this functionality. Its aim is to permit modelling at a high-level and to interconnect a range of common components from the synapse, neurons and plasticity models in whichever manner the user chooses. It is possible to think of PyNN as a common overlay that unites many simulation modules under a common description language. This is a true advantage when it comes to replicating experimental results.

- Nengo [STE09] is a tool which builds models using neuron populations (ensembles in Nengo terms) which are then interconnected to other ensembles. Each ensemble represents a value in the system, one which may alter as the simulation progresses, and each connection represents a calculation which occurs between the ensembles. The simplicity of the approach belies the complexity of models it is able to create, all of which rely on the underlying Network Engineering Framework (NEF) [EA03]. The NEF is a framework used to represent biological 'variables' encoded in the activity of recurrent neural networks, which permits control theory to be applied to spiking neural networks. The user is presented with a GUI in which they are able to use drag and drop techniques to construct their network, and then see visualisation of the operation of the networks.

- GENESIS (or GEneral NEural SImulation System) [BB07] is an environment which aims to provide an extensible simulation environment for biologically realistic neural networks. GENESIS has been designed to extend to parallel implementations in clustering or supercomputer environments, running large numbers of biologically realistic cell models. The users specify the network via a object-oriented scripting language which provides a GUI environment for the users to look at visualisation and simulation parameters.

- NEST (or NEural Simulation Tool) [GDG07] has been developed to create a simulation environment for large networks of neurons, concentrating on biologically realistic models with varied dynamical accuracies. This model is not suitable for users who wish to alter or examine neuron characteristics as these are abstracted during the compilation process. NEST uses scripts to specify and execute its networks, providing wide-ranging support on many operating platforms, and its output is provided in the form of data files which may be post-processed as required by Python or tools such as MATLAB [Mat12a]. Compatibility of the simulation engine with PyNN network descriptions, as noted earlier, is provided through the PyNEST binding layer.

**The Commodity PC and Workstations**

Whilst not specialised in any way for running neural networks, the hardware available in a workstation proves eminently capable of running small neural simulations, and fulfils many users' requirements. Users have the flexibility to choose to operate one of the software packages mentioned above, to customise them (many are open-source), or even write bespoke software to meet their modelling requirements. The prime disadvantage of the PC is that it is not readily scalable, so larger simulations may be unable to run or take longer to operate, even where the software is able to take advantage of any multi-core processing capability.

## Clustering Solutions

Several of the more popular neural network modelling software packages are capable of executing in parallel environments, and the logical expansion beyond a single workstation is to use multiple machines to expand the available computing resource. In neural simulation the inherent parallelism results in the smallest quantum typically being a single neuronal structure, and message passing is used to distribute data amongst these components. There are several techniques and software frameworks used to provide clustering facilities above the 'bare metal', and a message passing mechanism such as MPI [MPI09] is used to communicate amongst the computational nodes. Within the neural network realm, in 2007 Plesser et al. [PEM+07] used NEST to create a 12,500 LIF neuron network implementation operating on Sun cluster environments utilising a variant of MPI. Izhikevich and Edelman published work [IE08] in 2008 on modelling a much larger-scale mammalian cortex simulation with a million neurons and half a billion synapses at $\frac{1}{60}^{th}$ of real-time on a 60 processor Beowulf [SBS+95] cluster machine.

Unfortunately cluster computers may not always be suitable candidates for simulating neural networks, particularly those operating in real-time, as message passing over the interconnection network may introduce undesired latency. Delays due to latency in real-time systems form a pinch-point, becoming part of the computational cycle (in a synchronous network), or bounding the performance in asynchronous simulations.

A variant of clustering, where the nodes are less likely to be homogeneous, co-located, or even within the same management or ownership domain is grid computing. Again this appears to be an appealing solution, making use of processing cycles across workstations to create a high-throughput large computing resource [Fos03]. Grid solutions typically function by dividing the workload into coarse-grained parcels of work, but by their very nature may process these parcels using spare cycles (e.g. Condor [LLM88]). This method therefore appears unsuitable for neural network simulation which is usually latency bound. It has not been possible to find significant publications using grid techniques for neural network simulation, although for non-time critical applications grid computing may be suitable for parameter-sweep type experiments, where lots of smaller experiments occur independently and the results are collated and analysed later. Although not simulating artificial neural networks, grid computing has been used recently for neuroscience activity, for example the 2005 NeuroGrid initiative [GLS+05] and in the Ukrainian National Grid [SLS11].

**Supercomputers**

For larger scale networks, general-purpose, high-performance computer (HPC) platforms (supercomputers) may be used. Neural networks modelled on (IBM Blue Gene) HPCs include the Blue Brain [Mar06] initiative, which uses software which is a logical extension of both NEURON [HC97] and NeoCortical Simulator (NCS) [Bra12a] software to perform high fidelity cortical simulation work. The Blue Brain project is proposed for continuation as the Human Brain Project [Uni12]. Early work performed in SyNAPSE [AESM09] involving contentious claims of modelling at a scale of the cat cortex was also performed on this IBM supercomputer architecture. In software NEURON [HC97], in particular, advertises its support of HPC, noting particular compatibility with the IBM Blue Gene family of supercomputers [KHH+09]. Supercomputers are typically good environments in which to undertake large simulations due to their high concentration of homogeneous computing resource, and the well-provisioned interconnection networks.

**Use of GPUs**

Accelerated architectures such as GPUs [FH08] appear inherently suitable to the task of neural network simulation with their SIMD-type architectures, and are available from a number of manufactures including AMD (OpenCL) and nVidia (CUDA). Applying general purpose applications to graphics processors (GPGPU) for the implementation of neural networks has been explored in many research papers, [PBS11, HT10, BPS10, BPS10, BK06], typically reporting a many times speed-up compared with single threaded general CPU implementations. The great challenge with such implementations is keeping up with the memory requirements of the GPUs and in intercommunications. Software platforms have been created around the application of neural network modelling problems to the GPGPU architectures such as NeMo [FRSL09] which concentrates on the real-time simulation of hundreds of thousands of spiking neurons in real-time, and similarly GPU-SNN [NDK+09]. Whilst GPUs with their massive parallelism seem to be ideal for neural network simulation, there are issues with the scaling at points of serial operation such as synchronisation (Amhdal's Law [Amd67]), and the power-performance of GPUs compared with more attuned platform choices. Exaggerated claims of the speed-up possible using GPGPUs led researchers from Intel to write a paper (albeit not without vested interest) to debunk some of the more unrealistic comparisons [LKC+10], typically where the equivalent CPU code had

## 2.3. SYSTEMS FOR NEURAL NETWORK MODELLING

not been optimised. Until recently GPGPUs have been deployed in (and thus limited to) single host systems but are now found in larger scale computation platforms such as clusters [HT10], and in supercomputing deployed as co-processors to computing nodes and permeate the Top500 list [MSDS12]. These hybrid deployments are beginning to overhaul some of the scalability issues surrounding interconnectivity and maximum deployment size which have thus far limited GPGPU ANN simulations.

### 2.3.2 Specialised Hardware Platforms

For non-specialised platforms, bottlenecks in ANN simulation typically occur as they encounter communications pinch-points; large neural networks tend to generate heavy amounts of interconnecting traffic, particularly as simulation moves closer to 'real-time'. This leads to the requirement to build specialised solutions for large scale ANN implementation [IE08].

**Neuromorphic Hardware**

Neuromorphic hardware [Mea90] fuses analogue electronics and very large scale integration (VLSI) electronic techniques (perhaps in combination with some digital components) to create real-time bio-inspired circuitry. As neuromorphic components are modelled directly in hardware they are well placed to model spiking ANNs with far greater efficiency than general purpose computing. Although the lack of reconfigurability in neuromorphic hardware may seem limiting, the advantages of providing fast, power-efficient, solutions for neural network modelling may balance out the restrictions. However, any advantage of neuromorphic hardware may be eroded over time as the speed of development in general-purpose computing hardware tends to negate any initial advantage within a few years. Neuromorphic building blocks include silicon neurons and control components, and this modular nature permits efficient implementation of large neuromorphic systems which mimic the operation of biological circuits. Meade's implementation of a silicon retina spawned neuromorphic engineering [Mea89], and silicon retinas have been popularly developed by different labs: [CECB03, LPD08, CSSGSGLB07]. Liu and Wang's survey paper [LW09] covers most major neuromorphic hardware and initiatives. The efficiency of neuromorphic hardware leads to the opportunity for implementation of large-scale real-time biologically plausible networks [ILBH+11]. Larger scale neuromorphic systems have been created to form powerful computational platforms including Neurogrid and FACETS:

Boahen's *Neurogrid* project [Boa06] is a neuromorphic architecture that uses a custom analogue chip (a neurocore) to emulate up to $2^{16}$ neurons per chip. Each board contains 16 neurocores and provides a digital communication grid infrastructure known as softwires [MASB07] to interconnect the neurons with routed address event representation (AER) rather than providing discreet paths for each connection. Each chip supports up to 60,000 spikes per second to downstream synapses operating at ion-channel simulation level, with an aggregate 1 million neurons per board and billions of synapses, consuming just 5 W of power. The aim of Neurogrid is to produce an affordable supercomputer that operates in real-time for a fraction of the price of a general purpose supercomputer, albeit at the cost of losing some reconfigurability.

*FACETS* (Fast Analog Computing with Emergent Transient States) and latterly *BrainScaleS* [Bra12b] are European projects aiming to exploit biologically observed behaviours in novel computational environments. PyNN [DBE+09] was developed as part of this project to help standardise network descriptions. On the neuromorphic front the project has examined wafer-scale integration to exploit the communication density offered by interconnection of many constituent HICANN analogue chips by digital channels and crossbar switches on 20 cm diameter wafers [SFM08]. By operating in this wafer-scale environment it is anticipated that communication, which is typically the bottleneck in neural network simulations of any biologically significant scale can operate effectively with simulations running $10^4$ times faster than real-time. This is much faster than could feasibly be achieved in a reconfigurable platform for ANNs of equivalent size.

Finally in this section the second stage initiatives of the *SyNAPSE* project [Def12] propose the specialised Cog Ex Machina platform, which makes use of memristors [SAC+11]. In the classical model of neural network modelling, processing is applied in each and every case to work out how a post-synaptic neuron is impacted by a spike arriving at one of its synapses (the weighting typically being recovered from memory which is onerous). If in simulation this operation is replaced by the application of a memristor (which maintains its state between operations), then the spike data may be moved and manipulated (by the memristor weight) without reference to central processing or memory – just as if distributed in the brain through the dentritic tree. The machine is expected to be constructed from 'Dendra' chips which are comprised from two types of processor which act as transforming engines for data passing through them [VC10]. Firstly a neuron type processor (as there is in ANN simulations today), and the second 'dendritic' processor which acts as described above, formed of

## 2.3. SYSTEMS FOR NEURAL NETWORK MODELLING

memristors. Overlying this system and its components is a software layer known as 'cog' which abstracts the hardware implementation away from the users, who can concentrate on the functionality, with the operating system MoNETA (Modular Neural Exploring Traveling Agent).

**FPGA Accelerators**

Field-Programmable Gate Arrays (FPGAs) are used in a number of neural network simulations, due their advantages of being reconfigurable and providing fast operational performance. One of the key disadvantages of the FPGA is perceived as the expense to acquire the hardware, particularly in high-end devices, the costs scaling linearly with the number of devices deployed. This contrasts with the development of custom hardware where typically the expense is front-loaded and the incremental price is small. In building a simulator where the choice is between FPGA and custom silicon, scrutiny of the costs will determine the financial cross-over point between each approach. One further issue with FPGAs is that they are orders of magnitude less energy efficient than custom hardware [PPM+07]. Historically FPGAs have been used to prototype, and as a stepping-stone to dedicated hardware solutions.

In the work performed by Rice et al. [RTV09] a hybrid supercomputer, a Cray XD1, is used as the target platform; this has available 'hardware acceleration' in the form of 144 FPGAs augmenting 432 dual-core AMD Opteron chips. When examining speed-ups reported a cautious approach needs to be taken with the GPGPU figures [LKC+10], as it is not certain whether the native performances have been properly optimised. In this case the study reports a 75x speed-up due to the parallelisation of the model made possible by the FPGA hardware acceleration resource. Within the 2007 study performed by Pearson et al. [PPM+07], both connectivity and power problems were found to be issues when using FPGAs for real-time simulation of LIF-type neurons controlling a closed-loop task. While the connectivity problem could readily be resolved at the price of fewer simulated neurons per constituent FPGA, the energy efficiency problem was not surmountable in that generation of FPGA hardware.

FPGAs, like GPGPUs have deficits in communications when dealing with the unusually heavy bandwidth requirements of artificial neural networks. This may result in the neuron density of an FPGA solution being artificially constrained by the communication overheads [HMH+08]. With contemporary (2011) hardware it has been demonstrated that 1 million neurons are capable of being simulated in real-time [CAG11], and Moore et al. [MFM+12] propose 4 million real-time Izhikevich [Izh03] neurons on their Bluehive architecture fabricated using standard FPGA development boards and custom interconnection fabric (12 full-duplex 6 Gb / s SATA links per board). While FPGAs have the advantage of being programmable, their cost and power consumption limits their ultimate scalability [PPM+07].

**FPAA**

Field-Programmable Analogue Arrays (FPAAs) act in a similar manner to an FPGA in that they are reprogrammable and allow the user to create circuits using modules contained within the FPAA itself, rather than fabricate new silicon. They are ideally suited for analogue sensors and actuators as no conversion between analogue and digital or sampling is necessary, making transformations in the analogue domain efficient. FPAA has the further advantage of operating in a low power signal processing space, making them of interest to neural modellers as they can effectively provide a form of programmable neuromorphic hardware. FPAAs have been investigated for use in spiking neural network simulation [RMM+08], and derivations to the Field-Programmable Neural Array (FPNA) latterly by Hesler et al. [FGH06, BRP+10], and at this early stage appear a promising alternative to FPGA and more traditional neuromorphic hardware in the neural modelling environment.

### 2.3.3 The Centre Ground

Summarising thus far, bespoke neuromorphic hardware appears the most energy efficient but remains generally inflexible in what it can model; the general purpose approaches, whilst having model flexibility, use more power, and may not be able to cope with the traffic patterns of neural network modelling. There is room for an approach that tackles both sets of issues and occupies the central ground – whose structure is optimised for neural computation, but remains a general purpose parallel programmable architecture.

**Connectionist Network Supercomputer**

One such example originates in the early 1990s a team at U.C. Berkeley worked on the Connectionist Network Supercomputer [ABF+94], a hybrid supercomputer specifically tailored for neural computation. The system was designed as a 2D mesh of 128 (possibly scaling to 512) nodes, each incorporating a general-purpose RISC processor plus a vector co-processor. Its architecture included a host machine, directly attached to the mesh interconnect, which relied entirely on spatial locality to scale to the required performance. A prototype of the node was built under the codename T0, but it is not believed the system operated as a large network. The results of experiments using up to five nodes in a bus configuration were published [PA97].

**SpiNNaker**

SpiNNaker is an event-driven platform which attempts to straddle the requirements of full programmability, but additionally provides a high-degree of connectivity for artificial neural network modelling. SpiNNaker aims to be at the energy-efficient end of general purpose programmable platforms by using standard ARM processors, and provides a specifically designed scalable network infrastructure to cope with the expected traffic load from a biologically-plausible real-time Spiking Neural Network (SNN). Due to its general purpose programmability, the models to be simulated can be chosen flexibly and deployed heterogeneously throughout the machine as desired, with the chosen level of biological fidelity. For those seeking to compartmentalise the hardware, SpiNNaker is a *neuromorphic* [Mea90] or *neuromimetic* [RJG$^+$10] approach, as the needs of the biological simulations drive the requirements of the constituent hardware, particularly in SpiNNaker's communications fabric, and event-driven nature. However, there is no specific neural hardware in the system – the neural aspect is all provided in the software, and SpiNNaker may, like a general purpose computer, be deployed for non-neural computational jobs.

Therefore SpiNNaker successfully occupies the middle ground of being a specialised multi-purpose platform, whose interconnect is tuned to the needs of a SNN simulator. Chapter 3 provides fuller coverage of the SpiNNaker architecture.

# Chapter 3

# The SpiNNaker Neural Architecture

The SpiNNaker initiative, led by the Universities of Manchester and Southampton, and partners, is to create a biologically inspired, massively-parallel, computing architecture optimised to simulate very large-scale **Spi**king **N**eural **N**etworks in real-time [FTB06b]. Whilst it may be adapted for other computational purposes, the requirements of neural computing have dominated its design, creating a novel application-specific high performance computing (HPC) architecture. SpiNNaker's philosophy is to achieve these design goals with frugal use of power and resilience to component failures within the system [PFT[+]07, FB09].

SpiNNaker's targets of fault-tolerance and minimising energy use mimic the characteristics of the biological brain which, in humans, achieves incredibly high performance using billions of simple, fundamental, processing components (neurons) working in parallel [WH88, Dow01]. Biological neurons are slow, but highly-interconnected units; electronic components are fast but have much lower fanout. The SpiNNaker design therefore trades-off these properties using processors to simulate neuron and synaptic behaviour, and a fast network to deliver communications over packet-switched inter-connections.

Modelling full or fractional 'brain-sized' networks in real-time requires a huge number of processors, with the network capacity to match. SpiNNaker is designed to be expandable to biologically-significant sizes [HH09], and with large numbers of processors comes significant amounts of data.

## 3.1 Architectural Requirements

As far as is understood, biological neurons communicate primarily using electrical impulses known as spikes; each spike is a 'digital' signal in that it is either present or not. Output variations are represented by the *temporal* spacing of spikes [Izh06]. Other connectivity information is in the *synaptic weight,* which indicates how strongly a spike affects each post-synaptic neuron. Simulation of spiking neural networks is known as the third generation of artificial neural networks [Maa96].

SpiNNaker represents a spike as a single, short communications packet, which is *multicast* into the communications network where it may be replicated to a preprogrammed – possibly large – set of destinations. Typically, in a biological neural network, the input and output connection 'fan' of a neuron may be of the order 1,000 to 10,000 and sometimes up to 250,000 [NLMA+09]. This electronic transmission of a spike is nearly 'instantaneous' compared with biological timescales [Boa98] and the 'real' biological axonal / synaptic delays are modelled by the receiving processors in software [RJKF09, JGP+10].

Depending on the neuron model used [JGP+10, RGJF10, RGD+11], a million processor SpiNNaker system would support around 1 billion neurons in real-time, thus connectivity may exceed 1 trillion synapses. At an expected biological firing-rate of 10 Hz [DA01] there could be 10 billion-plus neuron firings per second which amplify in the output fans to trillions of communication events / s. SpiNNaker's network fabric has been scaled assuming significant locality of spike traffic (destination neurons are statistically proximate to the transmitting neuron), as is typically seen in the brain [BDM07]. SpiNNaker therefore has been designed to distribute huge numbers of short packets very widely amongst hundreds of thousands of processors efficiently, and in a timely fashion.

## 3.2 System Architecture

To create a high-performance processing and interconnection environment, the SpiNNaker architecture is constructed using custom designed Multi-Processor Systems-on-Chips (MPSoCs). An MPSoC is similar to the previously defined CMP, but includes additional components to form a self-contained system on that chip. Each SpiNNaker MPSoC (fig. 3.1) contains eighteen ARM9 cores for processing, and for connectivity

52                CHAPTER 3. THE SPINNAKER NEURAL ARCHITECTURE

(a) A SpiNNaker chip and stacked SDRAM (src: Unisem Europe)    (b) Packaged production SpiNNaker chip

Figure 3.1: Photographs of the SpiNNaker die, memory and package.

a bespoke router and communications fabric with full-duplex ports for the 18 internal cores and the six external chip connections. As spiking neural network models are essentially event driven environments (based on the arrival time of each spike), embedded-type processors such as the ARM are well suited as the cores remain dormant and in a low-power mode for a significant proportion of the time. SpiNNaker chips (nodes) may be interconnected with their six neighbours using a 2D triangular mesh wrapped into a torus (fig. 1.1). While not biologically realistic, this was chosen as an extensible configuration with a wide choice of redundant chip-to-chip routing paths providing high aggregate bisectional bandwidth.

Figure 3.2 is a simplified schematic overview of the SpiNNaker chip. Visible are the connections of each ARM core and external inter-chip link to the Communications Network-on-Chip (fig. 3.2: Comms NoC) – with the on- and off-chip networks forming a seamless routed whole. This is facilitated by an *asynchronous* interconnection medium so the *whole* is a GALS (Globally Asynchronous, Locally Synchronous) system [Cha84]. The asynchronous interconnect primarily provides better power economy than a synchronously-clocked alternative with the anticipated network loading patterns [EFEL05]. The asynchronous NoCs use the Chain [BF02] technology supplied by project partner Silistix.

A second GALS NoC (fig. 3.2: System NoC) gives all cores access to shared peripherals and a separate (in-package stacked) 128 MB SDRAM (fig. 3.1a). This network has quite different requirements: it has to supply sizable data blocks to the processors, usually under DMA control. A shared on-chip 32 kB SRAM – which is used for inter-core message passing communications – is also addressed using this path, as is the ROM which contains the boot software.

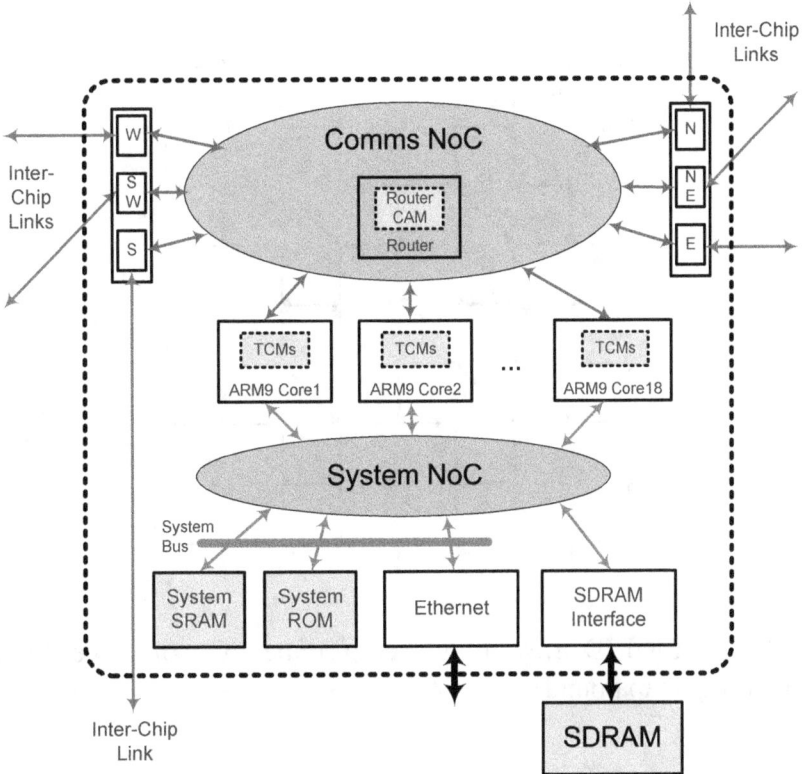

Figure 3.2: Simplified schematic of the SpiNNaker chip (node).

Input / Output is provided primarily by 100 Mb / s Fast Ethernet (figs. 1.1, 3.2) provisioned on a subset of chips. There are also some General Purpose I / O (GPIO) lines which may interface with external devices at a somewhat lower speed. Further details on the overall SpiNNaker system architecture and its design philosophy to tackle spiking neural computation problems may be found in the following publications [FT06, FTB06a, PFT+07, PBF+08, FB09].

## 3.3 SpiNNaker Systems

In its maximal configuration the SpiNNaker platform is designed to simulate around 1% of the neuron count of the human brain [FB09]. A fully configured SpiNNaker system achieves this by combining SpiNNaker chips into a massive single machine interconnected as a 6-way toroidal mesh (fig. 1.1). A square system, laid out and tiled two dimensionally, can range from single chip to $256 \times 256$ chip configurations – the edge connections wrapping around to modulo $x$ and $y$ opposites, as illustrated in figure 3.3.

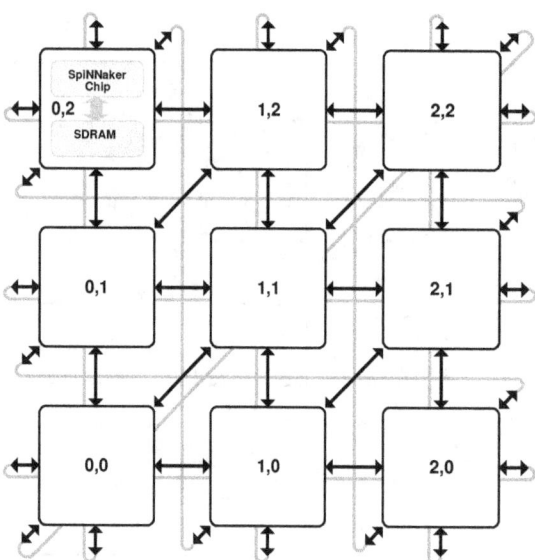

Figure 3.3: Regular tiled 2D array of multi-chip connectivity of a nine chip SpiNNaker machine (detailing wraparound).

Each ARM processor running at a nominal clock speed of 200 MHz is able to simulate around 1,000 Izhikevich neurons in software using the approach taken by Jin et al. [JFW08]. The assumption therefore is that around 16,000 neurons can be simulated per SpiNNaker chip (16 cores are earmarked per chip for simulation work). In larger systems this requires relative locality of input neurons and populations rather than a random distribution [NPMA+10]. This is biologically consistent with studies of the mammalian cortex, which is found to have deeply interconnected local structures, with sparse longer range connectivity [BDM04, BDM09]. It is expected that any one moment (using millisecond intervals [Tho90]), a sparse 5-10% input connectivity ratio is expected (spikes per interval per neuron input), with each neuron typically achieving a maximum firing-rate of around 60 Hz.

Figure 3.3 shows the 2D Cartesian co-ordinates given to each chip (the origin (0,0) can actually be arbitrarily chosen due to the wrap-around of connections on the 'edge' chips). Sixteen bits are allocated for addressing ($x$ and $y$ co-ordinates in the Cartesian case), therefore a maximum 65,536 chip system size is possible. One of each chip's 18 cores is elected to the rôle of 'Monitor Processor' (MP) which coordinates chip-level functions such as non-spike communications, control and management. To remain fault-tolerant, this assignment is made dynamically after power-on testing from the set of known-good processors. It is expected to deploy chips where either 18 *or* 17 cores are functional to increase yield, with 16 assigned as 'Application Processors' to perform the neural simulation work. The 18th core, where functional, is designated 'spare' and may be assigned to utility purposes, or left unused. Whilst this is not a faithful biological replication of the brain, it takes advantage of the electronic medium being used to create the neural network simulations, whilst maintaining the spirit of the *redundancy* inherent to biological systems. With 16 application cores for neural simulation, there are $2^{16} \times 16 = 1.049$ million cores in a full system, each potentially modelling a thousand Izhikevich-type neurons, giving the opportunity to create real-time simulations of over a billion neurons (around 1% of the neuron count in the human brain ([WH88]).

## 3.4 SpiNNaker Communications

### 3.4.1 Operation

Neural-network simulation is a highly parallel task; each core is loaded with neural processing software and its neurons' local synaptic weights. After initialisation of all chips, cores run asynchronously and handle events occurring in real-time within the system. Spike packets are routed and replicated in hardware through the machine's communication fabric into multicast trees.

It is anticipated that typical customers for SpiNNaker are psychologists, neuroscientists and multi-disciplinary teams creating neural network simulations using biological principles, as well as others bringing their highly parallelisable tasks to the platform. These users will wish to gain visibility of application software and hardware performance data 'in-flight', thus the communications fabric must also support transit of monitoring and debugging management traffic.

The requirements of management services are different from those of spike traffic, but to ensure neural traffic is not disrupted, the alternative packet types employed retain the same short message length principle. Longer messages and reliable delivery are handled by higher-level software protocols.

SpiNNaker, unlike many typical HPC architectures (section 2.3), has no separate out-of-band control and instrumentation channel. This choice is based on the relatively low duty-cycle targets of the interconnection network and to reduce MPSoC design complexity, and thus power, by operating a single, merged network. Therefore spike event packets and real-time management and diagnostic traffic share the same network and may experience contention if assumptions are exceeded and hot-spots occur. It is expected that only a small proportion of traffic traversing the networks will be used for system management and instrumentation.

SpiNNaker features interconnection networks at many scales and sizes, from on-chip to inter-chip, to ex-machine I / O connectivity. It is a marriage of various communication and encoding protocols: ARM's AMBA [ARM11], plus asynchronous on-chip 3-of-6 RTZ, chip-to-chip 2-of-7 NRZ and Ethernet. There is a layered approach to connectivity, with standard and bespoke protocols enabling high-performance neural messaging and system management functions to be efficiently supported over all physical transport paths on a large-scale machine.

On the Host system software functions are provided to allow diagnostics and monitoring of hardware and software to be turned on and off, so that the SpiNNaker machine does not operate its jobs as a 'black-box'. Monitoring functions will alert hardware operators to system issues, permitting the control software to terminate the simulations, or to dynamically remap around the problem. Other management-type functions include visualisation of the system – such as plotting fMRI type neural activity – monitoring resource utilisation (e.g. network links, processor cycles, memory) and bespoke neural parameter reporting.

### 3.4.2 SpiNNaker NoCs and the Router

Each SpiNNaker chip has two bespoke asynchronous Network on Chips (NoCs) which are highly efficient during periods of quiescence, as there is no continuous synchronous clock burning power.

The System NoC provides a path for cores to share chip-level resources, primarily conveying large quantities of data amongst its clients, including access to the shared SDRAM, System RAM / ROM and peripherals (fig. 3.2). Typically DMA is used to service high-bandwidth block data transfers across this NoC, but ad-hoc single-cycle, processor-originated accesses are also permitted – such as for message passing across shared System RAM.

The Comms NoC (fig. 3.2) is used to convey communications packets between cores and chips within a system, switched by the SpiNNaker router. The router has twenty-four asynchronous full-duplex ports: six for the external connections and eighteen for on-chip processors. The packet formats supported have been intentionally kept short as the primary use of the communication channels will be to convey spike traffic in a timely manner.

## 3.4. SPINNAKER COMMUNICATIONS

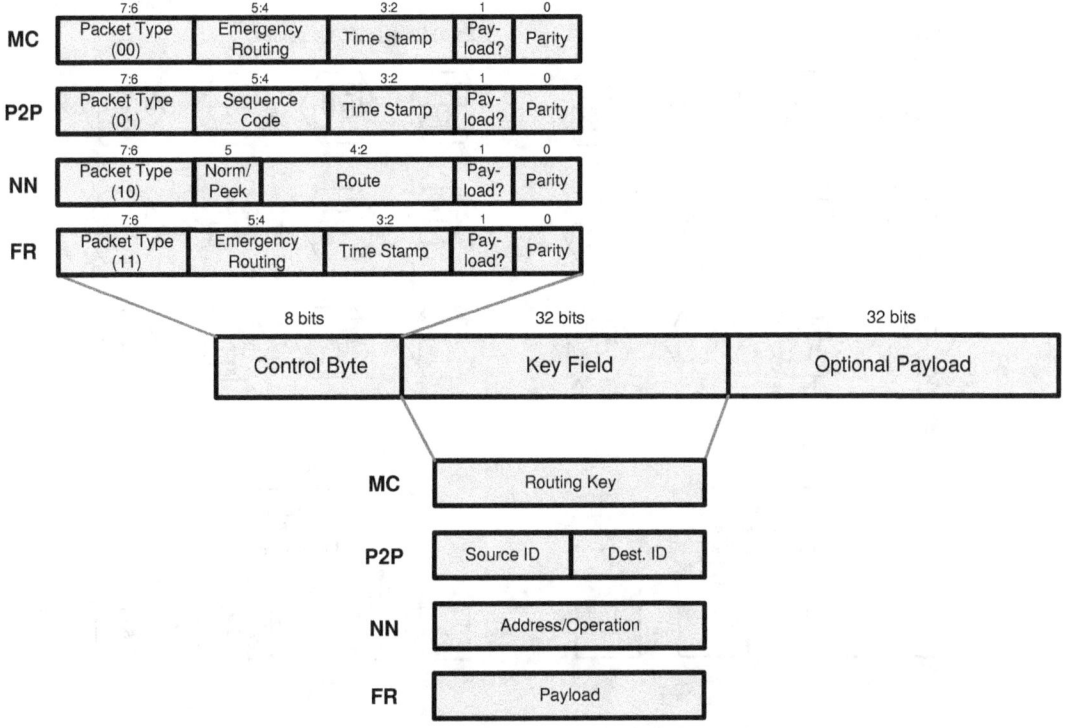

Figure 3.4: SpiNNaker packet formats.

### 3.4.3 Packet Formats

To cater for the different communication demands, four types of packet are supported by the router (fig. 3.4). Packets are either 5 or 9 bytes in length, comprising an 8-bit control header and a 32-bit field typically used for routing; which may be augmented with an optional 32-bit *payload* field. Example distribution trees can be seen in figures 3.5 and 3.6 for each of the four packet types:

- Multicast (MC): intended for neural spike events. *(one:many)*
- Point-to-Point (P2P): node-to-node communication, for code distribution and system control. *(one:one)*
- Nearest Neighbour (NN): principally for boot purposes and fault recovery. *(local node:node)*
- Fixed Route (FR): downloading data, analogous to a default route / gateway. *(typically many:one)*

The control byte (header) is similar for all packet types (fig. 3.4). Two bits identify the packet type, one bit indicates the presence of an optional payload and one bit

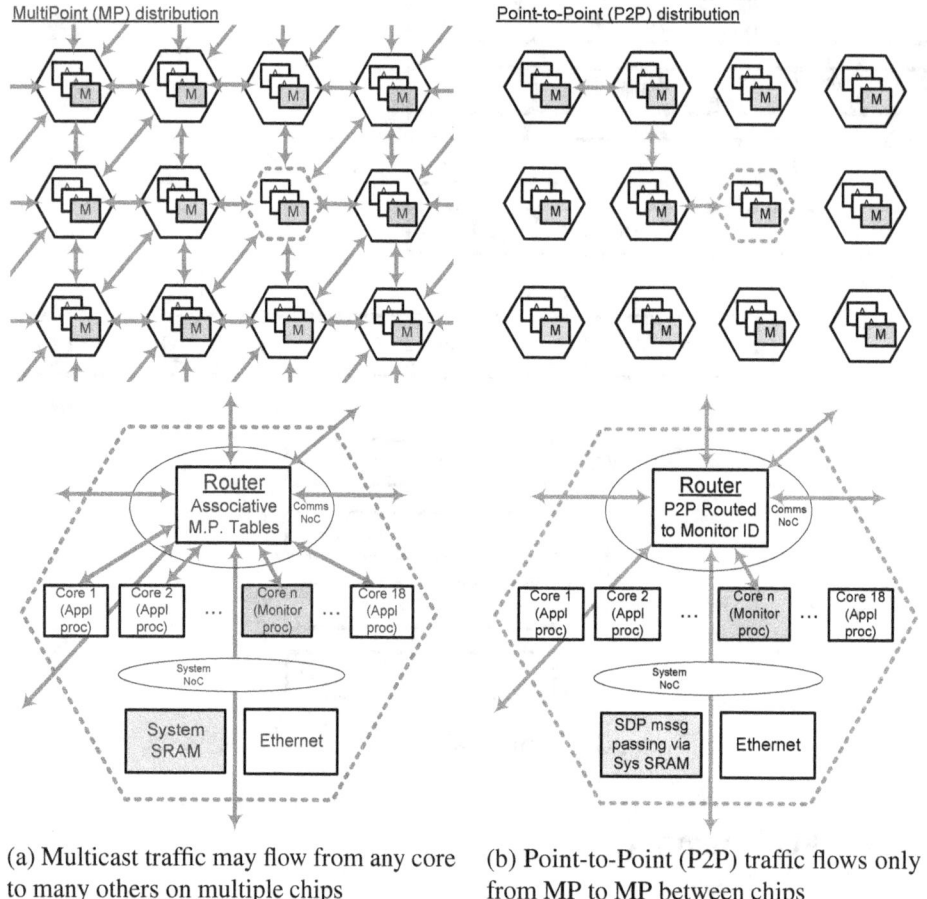

(a) Multicast traffic may flow from any core to many others on multiple chips

(b) Point-to-Point (P2P) traffic flows only from MP to MP between chips

Figure 3.5: Multicast and Point to Point packet flow and distribution trees.

records the entire packet's parity, including any payload. Most packets have a two-bit timestamp which allows routers to drop packets of a certain age, a means to filter 'rogue' looping traffic caused by faults. The MC and FR packets also have 2 bits of 'emergency routing' information to control routing around a failed or congested link [NLMA+09]. The sequence code field for the P2P packets facilitates the structuring of longer messages by higher-layer software protocols.

**Multicast Packets**

Multicast spike packets are distributed to a subset of the neural processors using Address Event Representation (AER) [MV94, Boa98, Boa00]. For this purpose each node contains an associative routing table consisting of a 1024-entry key, mask and target triplet. Where no match is made from this table the packet 'default routes' to egress opposite its ingress, meaning a table entry is only needed where a packet is steered

## 3.4. SPINNAKER COMMUNICATIONS

to destination processors, or where it needs to turn or bifurcate in transmission (e.g. fig. 3.5a). MC packets may be routed from and to every core in the system. In general they are intended purely as neural spike events and will not carry a payload, and are expected to dominate the network traffic.

**Point-to-Point Packets**

P2P packets target a single destination chip, not an individual core, and are delivered to the designated Monitor Processor (MP) on the chip. At each chip there is a P2P routing table which contains $2^{16}$ routing entries (1 entry for each node in a full SpiNNaker system). Each of these entries in the P2P table uses 3-bits to specify the direction of the next hop for that node, which is either 'here' or one of the six external links (fig. 3.5b). P2P packets are typically used for tasks such as code and data distribution, or management queries usually carrying a payload of higher software protocol layers and data.

**Nearest Neighbour Packets**

NN packets are used mainly as part of the boot process and for debug access to neighbouring chips. As the name implies, they are short range, permitting read / write access to a neighbouring chip's shared resources (fig. 3.6a). Their distribution tree is addressed by local link ID and not a routing table. The 'NN packet type' field in NN packets provides information to the router as to whether the packet is a normal packet (to be delivered to the Monitor Processor of the neighbouring chip) or a peek / poke packet (for fault handling). The route field enables an NN packet to be directed to the local Monitor Processor, a particular neighbour or broadcast to all or a subset of the neighbours.

**Fixed-Route Packets**

FR packets are similar to MC packets. The difference is that they are routed regardless of source by a single route-word at *each chip* so that only fixed, unidirectional merging tree structures can be implemented (fig. 3.6b). This packet type is designed to allow extraction by Ethernet of information at low cost both in routing hardware and bandwidth overhead (the 32-bit 'key' field is available for payload too). The routing entry may be dynamically altered, so, may form a useful distribution tree for loading and

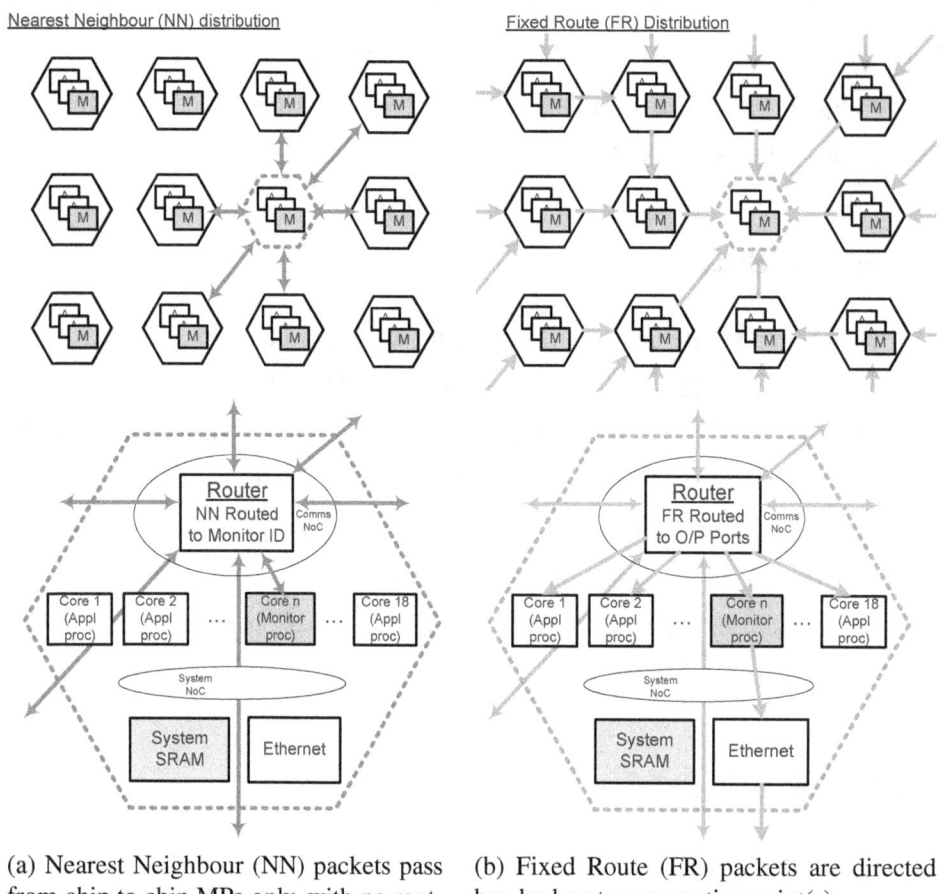

(a) Nearest Neighbour (NN) packets pass from chip to chip MPs only, with no routing

(b) Fixed Route (FR) packets are directed hop by hop to aggregation point(s)

Figure 3.6: Nearest Neighbour and Fixed Route packet flow and distribution trees.

saving chip specific data to and from SpiNNaker with the least overhead of any of the packet types.

### 3.4.4 Layered Networking

The *physical*, system-wide SpiNNaker network is optimised for expected traffic; primarily it presents a uniform 'flat' source-based routed medium for neural spikes to traverse to their destinations. The secondary load is machine-control and management traffic, usually from chip-to-chip, handled by the local Monitor Processors.

Provision for other communication needs (such as a Host probing an individual neuron model on some arbitrary core) is achieved by overlaying software protocols on top of the physical hardware and providing a hierarchical communication environment. There are four distinct identified interoperating communication layers:

## 3.4. SPINNAKER COMMUNICATIONS

1. Intra-chip – by the asynchronous Comms NoC and router in hardware. There are four types of packet as discussed in section 3.4.3. At this layer message passing across the System NoC using the System SRAM for core-to-core non-spike communications is also supported.

2. Inter-chip – the same packet formats, traversing between chips over the chip-to-chip external asynchronous links (fig. 3.3).

3. Ex-system – SpiNNaker Datagram Protocol (SDP) (section 3.4.7), developed within the group, provides for connectivity in and out of the machine via the Ethernet connections (fig. 3.7). In the near future FPGAs are expected to be interposed in the connectivity mesh to be used as additional ex-system paths.

4. Internet – beyond the local link, the same SDP software protocol is used to extend SpiNNaker communications in routed internetworks.

### 3.4.5 Ex-System, Ethernet

External I / O connectivity is provided by commodity 100 Mb / s 'Fast Ethernet' links attached to a subset of the SpiNNaker chips. Chips which detect the presence of a PHY (PHYsical layer transceiver) at power-on enable their Ethernet controller, otherwise it remains dormant to save power. The Ethernet is used to transfer neural spike information in and out of the system and for system control and management traffic. Traffic flowing via the Ethernet frames is constructed in software layers operating above the physical hardware and network layers.

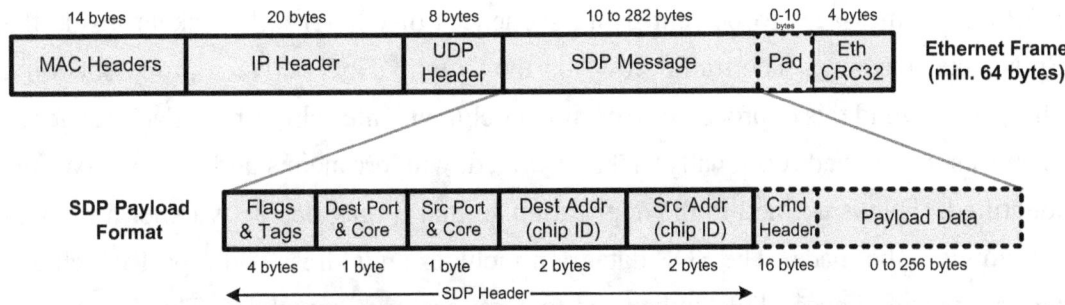

Figure 3.7: Simple SDP Ethernet framing format used for ex-system and Internet communications. Data is encapsulated by Ethernet, IP, UDP and SDP headers.

## 3.4.6 Internet Encapsulation

Internet Protocol (IP) encapsulation of SpiNNaker ex-system data permits remote collaboration to occur over routed networks including the Internet. Using IP facilitates remote access to the machine environment and enables standard 'sockets' programming libraries to be used to interface with the SpiNNaker communications stack.

Running a full TCP/IP stack on an ARM system can take 40 kB [Mic09], which would more than fill the entire instruction memory of the SpiNNaker processor. Hence only the fundamental requirements of IP and troubleshooting are implemented, including responders for Address Resolution Protocol (ARP) and Ping, as well as User Datagram Protocol (UDP), a low overhead transport mechanism. UDP was selected as a good match for neural spikes which are time-sensitive and 'one-shot', with no facilities for retransmission. UDP also requires relatively little implementation resource whilst providing the required functionality.

Transitioning between external and internal machine packets can be inefficient as a neural spike datum is small (4 or 8 bytes). To convey such an event to a Host device it must be encapsulated in an Ethernet frame which has a minimum frame size of 64 bytes (fig. 3.7), with the inefficiency being countered by aggregation of data.

All the layers as described above are *connectionless*, that is they do not store state or make any attempt to detect and retransmit data lost in the transmission process. The SDP protocol operates across all the layers of physical communication, from external Host to internal processors (fig. 3.8), and may carry data regardless of type including real-time I / O stimulus, program code, application data and management traffic.

## 3.4.7 SpiNNaker Datagram Protocol (SDP)

### SDP Internally Within SpiNNaker

SDP allows messages to be sent using sequences of (short) P2P packets inside the SpiNNaker machine. The traffic flows on the Comms NoC between processors on a chip, and beyond this to processors on another chip via inter-chip links. Each sequence is checksummed and (optionally) acknowledged, with erroneous and dropped packets identified. This is notified to the application so that it may decide whether a retransmission is to be made. The SDP datagram includes an address and a port which can be on *any* core; hence SDP can be used to pass messages anywhere. This is achieved by a chip's Monitor Processor relaying data to target Application Processors via the on-chip shared System RAM across the System NoC.

## 3.5. SOFTWARE ON SPINNAKER

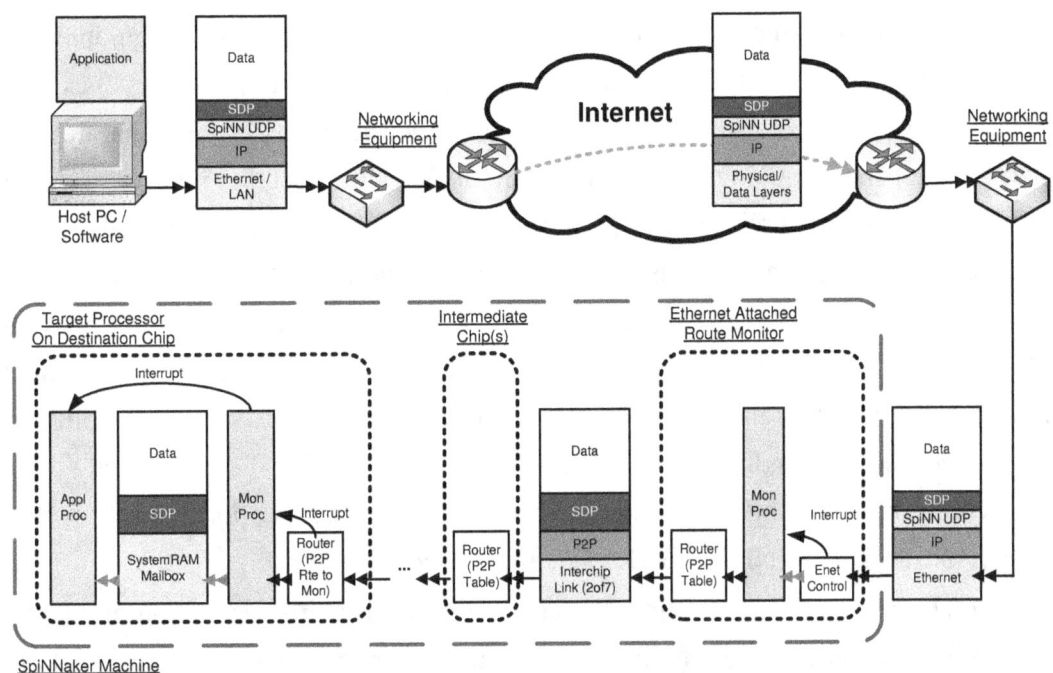

Figure 3.8: Communication layering diagram. Data flows from the Host application (top left) across all 4 distinct communication layers to the destination SpiNNaker Application Processor (bottom left).

**SDP Outside SpiNNaker**

A comparable mechanism is used for external communications where a Monitor Processor on an Ethernet-attached chip bridges P2P SDP packets into Ethernet SDP frames (albeit with fewer fragments due to the larger available payload (fig. 3.7)). A transfer from an external Host device to an internal processor target is depicted in figure 3.8, where the Ethernet-attached SpiNNaker chip acts as a seamless bridge between the internal P2P and the external Ethernet / IP domains for the SDP transport protocol.

## 3.5 Software on SpiNNaker

At the outset, SpiNNaker neural network implementations were hard coded into binary files that were loaded to the hardware [JFW08, RGJF10] providing proof-of-concept validations of particular models. This approach, however, is not scalable or generalisable. As SpiNNaker has the goal of providing reconfigurability in large scale systems, the group has developed a new approach of modular and descriptive modelling – which also abstracts the neural network description from the underlying hardware.

When creating artificial neural networks models which run on reconfigurable platforms, modellers tend not to work at the individual neuron and synapse level as there are such large numbers in larger simulations. Therefore sets of neurons are bundled together into groupings known as populations (mimicking the biological microarchitecture and assemblies ([JL07]), with populations being interconnected via projections which describe the statistical relationship of individual population to population interconnection. By abstracting at this level, large biologically plausible artificial neural networks may be created scalably and without direct reference to the target architecture. To apply the model description to the target hardware, a 'compilation' stage occurs which handles the mapping process. Section 2.3.1 discusses many of the most common modelling software packages, and SpiNNaker supports several including PyNN [GRDF10], NEST [GDR+12], and Nengo [GDF+12].

The abstraction technique aids the resource allocation process, as the problem is simplified in the initial stages to high-level granularity rather than to low-level cellular components. Within SpiNNaker this function is performed by PACMAN – the PArtitioning and Configuration MANager (fig. 3.9), which takes the high-level representation and transforms it into a SpiNNaker machine specific set of binaries [GDR+12].

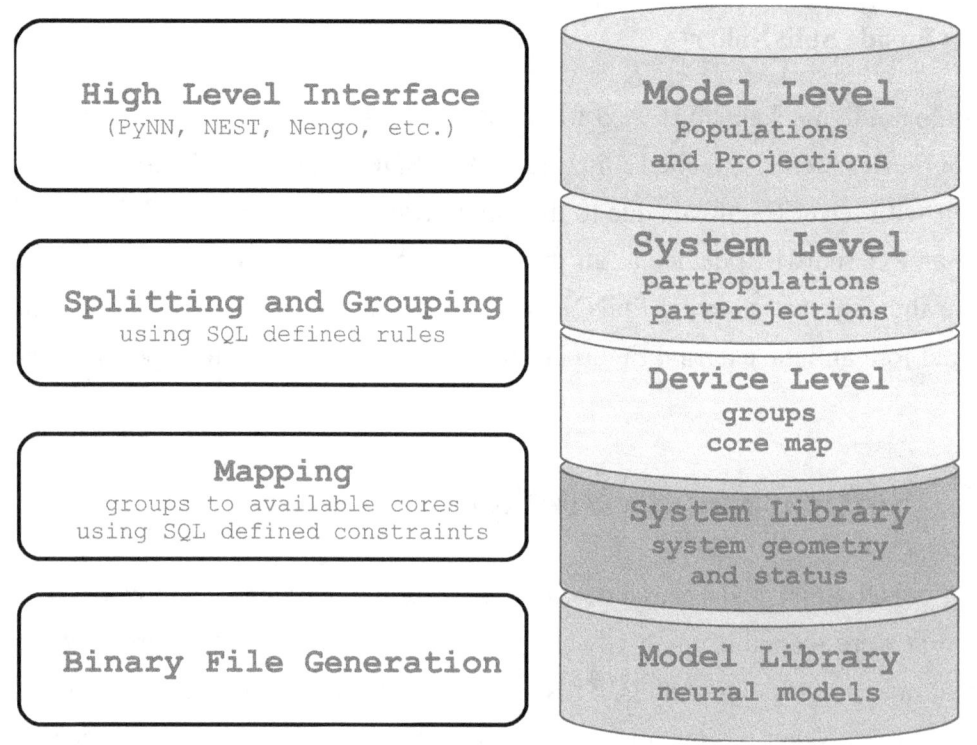

Figure 3.9: Partition And Configuration MANager (PACMAN) (src: [GDR+12]).

## 3.5. SOFTWARE ON SPINNAKER

PACMAN relies on several database representations of the neural network (top right of figure 3.9) which hierarchically decompose the original model level representation to the device level. The first stage is the splitting stage, where populations too large to fit on a single SpiNNaker core are split into part-populations, (together with their required connectivity). Next they are grouped onto cores, such that (part) populations may be placed alongside others in a group to maximise use of resources. At all stages of PACMAN SpiNNaker specific description libraries are used to inform the process.

The model library contains information about components used in the simulation, including the implementation resources (computational requirements / memory) required by particular neuron, synaptic and plasticity models. The system library contains the numbers and dimension information about the target system itself, which may be dynamically informed by management information about the health of the specific target machine. These libraries enable the device-level database to be mapped onto the specific hardware of the target system, including generation of the routing required to map between them all. It is this final hierarchical mapping stage which is used to generate SpiNNaker specific binary files, with neural network information and data tables to be loaded to each individual SpiNNaker chip and core.

At the small scale this place-and-route task is currently performed by a PACMAN instance on the external Host. In the longer term there are two major scalability issues to this approach:

1. Volume of Data. Although the executable code may be identical for each Application Processor (or be a small set of variants), the synaptic data set for each chip is unique, and this generated data set is *much* larger than the code. Serial loading of $\sim 8$ MB to $10^6$ application cores is clearly infeasible: with a single 100 Mb / s Ethernet link the load time would be measurable in *days*. More links can be added to reduce the transmission time, and perhaps advantage taken of the interposed FPGA connections in large systems.

2. A larger problem is faced by the Host performing the PACMAN task as the neuron place-and-route of a very large uniformly connected neural network is a compute-intensive NP-complete problem [GJ79, Bok81]. Although grouping the neurons into populations and the connectivity into projections is a viable technique for initial representation, connectivity is the predominating problem once the system is decomposed into its individual components [GRDF10]. PACMAN has therefore been constructed so that the latter stages of the mapping and

binary generation process may be parallelised. It is feasible that the aggregated smaller data sets can be initially distributed coarsely by the Host system with SpiNNaker itself performing the detailed generation of synaptic structures. This approach, planned for future PACMAN evolutions, impacts positively on the first limitation; less data is required to be transmitted into the system at startup and on the second point it exploits the distributed parallelism of the machine itself.

## 3.6 Summary

This chapter has outlined many of the architectural features of the SpiNNaker chip and its planned deployment in a million processor machine. SpiNNaker has been developed with the primary goal of contributing to the scientific Grand Challenge of achieving an understanding of the principles of operation of information processing in the brain. It aims to achieve this in hardware through the deployment of a specialised high-performance computer, and in software by supporting the execution of a wide range of simulations of artificial neural networks – scaling up to a billion biologically-plausible spiking neurons in real-time. The main innovation within the SpiNNaker architecture is the provision of a novel energy-efficient asynchronous interconnection network which has the ability to source and replicate very large numbers of small, independent, multicast packets each second. Together with the large numbers of standardised, but energy-efficient, programmable processors; this naturally raises the question as to whether the SpiNNaker configuration may benefit applications beyond the neural modelling space.

A machine of such a massive scale comprises many tens of thousands of components and, so, must also be designed with a resilient philosophy to tolerate faulty components rather than have the system repeatedly fail and require maintenance. The external Ethernet connectivity is provided for Input / Output stimuli and, crucially, for management purposes. This makes it possible to alert system operators of any alarms in the system and to enable them to be located and fixed or worked around. In the next chapter the first step of the SpiNNaker system management time-line is described and is the first major contribution to this thesis. Chapter 4 details the SpiNNaker boot process after power-up or reset, its testing, recovery and reporting mechanisms – getting a SpiNNaker machine to a managed state where applications can be run.

# Chapter 4

# Bootstrapping SpiNNaker

This chapter covers the boot process of the SpiNNaker chip following reset or power on and the areas in the system management time-line indicated in figure 4.1.

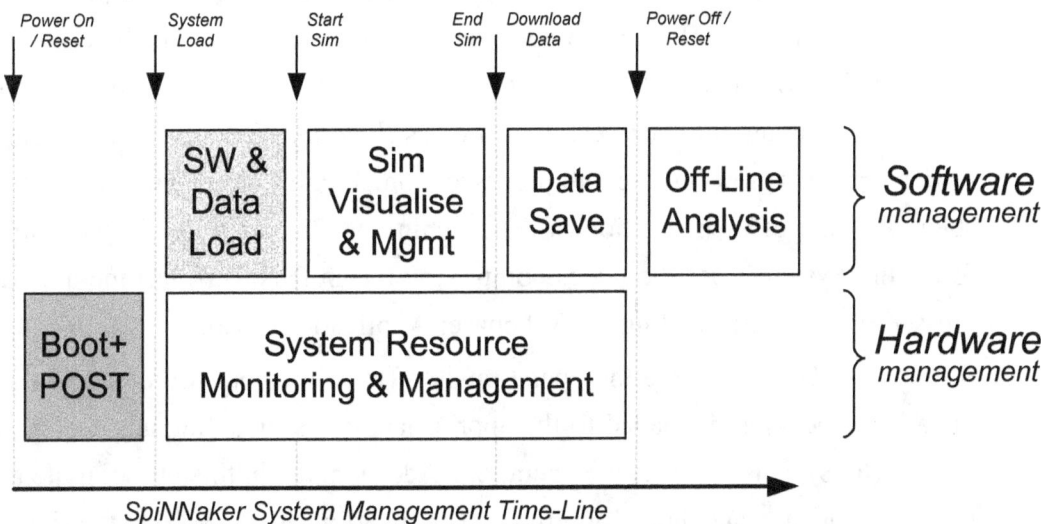

Figure 4.1: The management time-line of the SpiNNaker platform – The Node-Boot ROM tackles the POST manageability and supports the initial software and data load.

The boot process of a SpiNNaker machine may be summarised in three distinct phases (which are also depicted in figure 4.2):

1. *Node-Boot* is executed at power-on, with the Node-Boot code retrieved from the read-only *ROM* on each chip (node). This code performs the primary chip testing and initialisation, then handles election of a Monitor Processor and leaves the node ready to receive the externally originated 2nd phase System-Boot image.

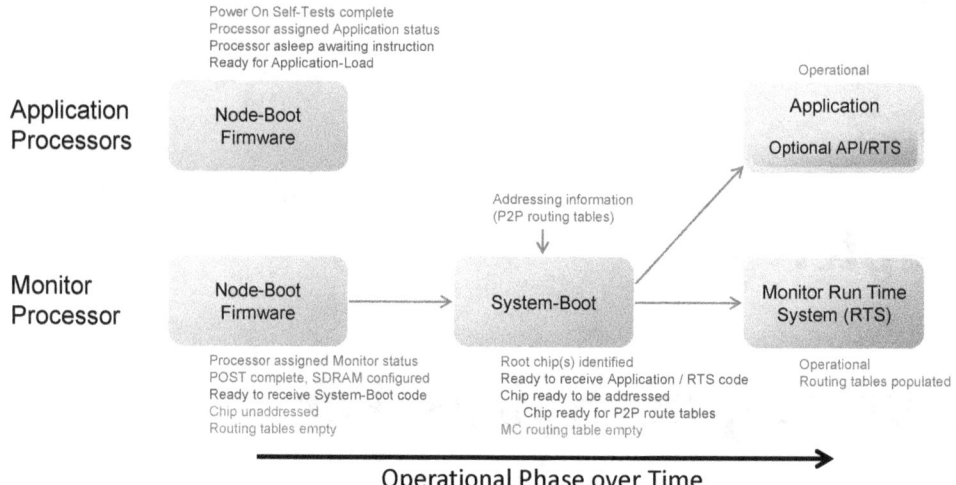

Figure 4.2: SpiNNaker boot and application loading sequence.

2. The *System-Boot* image is received by one or more Ethernet-attached SpiNNaker 'root' chips from the Host. The code self-propagates to its immediate neighbours, and so on, until all Monitor Processors in the system are running the System-Boot image; this process is known as *flood-fill*. Each SpiNNaker chip is identical so, following reset, all processors on all chips are in a homogeneous state, unaware of their position in the machine or their identity. Immediately following System-Boot the nodes comprising the SpiNNaker machine are addressed (numbered) enabling the 3rd phase: Application-Load.

3. *Application-Load* is where the operational software for both Monitor and Application Processors is loaded to the appropriate cores including any real-time system (RTS) that underpins its operation. Additionally, in this phase, route tables are populated and data supporting the applications is uploaded to the shared SDRAM of each chip. The application is then executed.

This remainder of this chapter covers the major design decisions and novelty contained within the 'ROM' (as *Node-Boot* software), particularly concentrating on those tasks related to the management and health of the SpiNNaker system.

## 4.1 Node-Boot

Node-Boot is the first stage of the SpiNNaker boot-up process (fig. 4.2) and is executed on power-up or reset by running code from the ROM (highlighted centre right,

## 4.1. NODE-BOOT

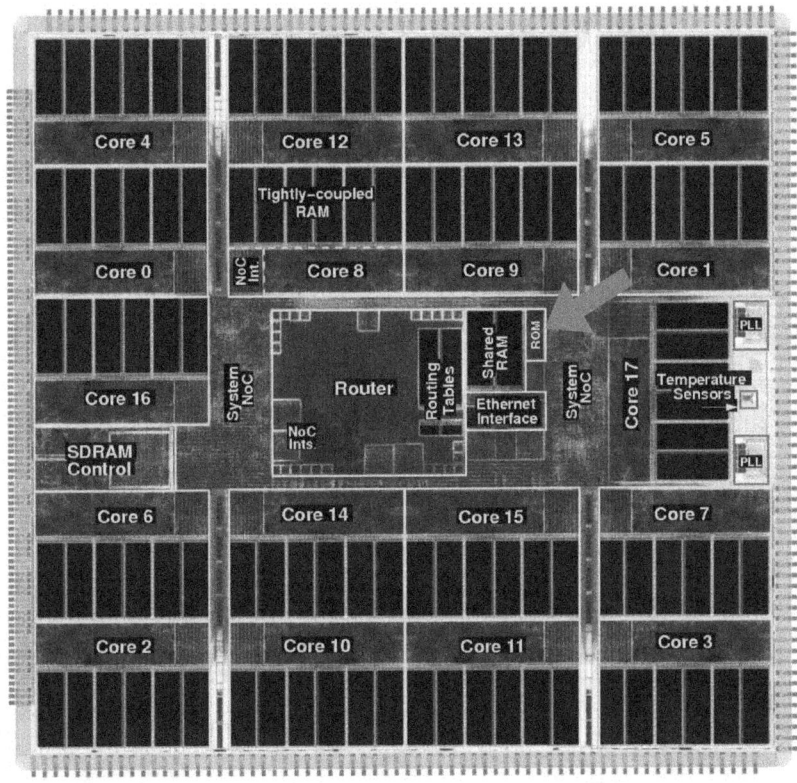

Figure 4.3: GDS2 plot of the SpiNNaker die (src: IMEC), augmented with labelling of each core and other major components (including the highlighted ROM).

figure 4.3). The Node-Boot code is executed by all processors on a chip for the primary purposes of performing Power-On Self-Tests (POST), hardware initialisation of components and peripherals, and to elect a Monitor Processor for the chip. Gracefully managing faults ensures that it may not be an immediate requirement to change a circuit board (which may contain dozens of functional chips), or cause downtime for the machine as a whole in the event of a component failure.

All 18 cores in each SpiNNaker node populate their instruction pipelines from the 'Boot' ROM. Once a functional processor passes its testing it enters 'listening' mode, awaiting a subsequent System-Boot software image to be downloaded to it. The System-Boot software, as it is external software, can be modified as necessary to add features and perform debugging as required – unlike the ROM content which is immutable.

The following sections outline the motivation, implementation and program flow of the boot sequence in greater detail. Refer to the two Node-Boot flow-charts (fig. 4.4 and 4.5), as necessary while reviewing this section.

# CHAPTER 4. BOOTSTRAPPING SPINNAKER

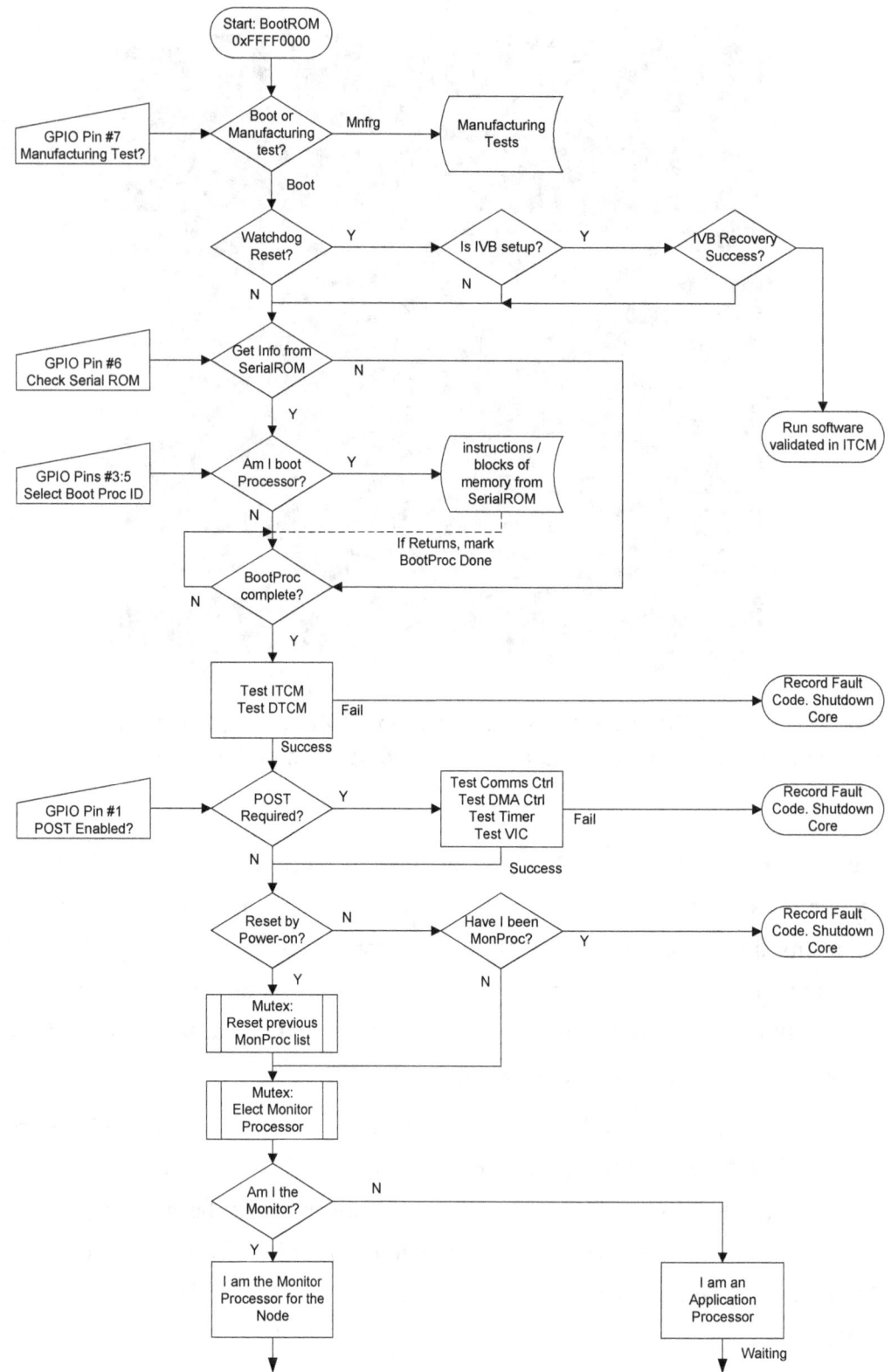

Figure 4.4: Flow of all cores during Node-Boot, from reset until Monitor Processor allocation. This flowchart continues in figure 4.5.

## 4.1. NODE-BOOT

### 4.1.1 First steps of Node-Boot

The first test in the boot sequence checks the hardware status of an external general purpose Input / Output (GPIO) pin (attached to the chip package) to determine whether a manufacturing test should be run (for die testing), or whether the default Node-Boot code should operate. Assuming the default latter behaviour, each core checks the chip-level 'Reset Code' register of the System Controller which influences the booting behaviour. Usually Node-Boot is encountered at power-on, but it may also be executed again if a reset is triggered, or if a watchdog enforced reset has been initiated due to errant run-time behaviour.

**Watchdog caused Boot** If the cause of boot is errant software behaviour triggering the watchdog timer, application authors may optionally provide a user-defined software restart mechanism which avoids passing back into Node-Boot. This novel facility (detailed in section 4.3) is called the ITCM Validation Block (IVB) which, if required, validates whether the instruction memory remains intact and uncorrupted by the software malfunction. If so, it branches to restart the core at a user-defined software function. If the facility is not used, or validation fails, the usual reset procedures apply.

**Usage of the Serial ROM** With non-watchdog reset codes, including power on or remote reset, the usual Node-Boot routines are followed to initialise and check the hardware. One of the cores is selected as the *Boot Processor* which checks for the presence of an external Serial ROM chip. The Serial ROM chip is used primarily to furnish Ethernet-attached chips with unique network addressing information (table 4.1), but may also, optionally, be used to exit the Node-Boot sequence early, to load and execute an image provided on the Serial ROM chip. One common use of this technique is to load a variant of the Node-Boot software which contains software to negotiate an address on a network using Dynamic Host Control Protocol (DHCP).

Whilst this occurs all non Boot Processors wait for a signal from the Boot Processor (on completion of the Serial ROM routine), before rejoining the main flow. Clocks for the cores, memory, router and system clocks are increased beyond the initial 10 MHz to safe, uprated, speeds to accelerate the remainder of the boot process. The per-processor Failure Logs are now cleared (table 4.1) so that hardware faults discovered during the power-on self-tests can be recorded, before the POST begins:

| | |
|---|---|
| 8 Words | Ethernet Parameters (loaded from optional Serial ROM) |
| 4 Words | Mailbox (message passing for image loading) |
| 2 Words | SDRAM Information (detected size, and any errors found) |
| 18 Words | Processor Failure Log (18 cores) |
| 1 Word | Monitor History (cores elected MP since power-on) |
| 257 Words | Shared Assembly Block (1kB + 4Byte CRC for flood-filling) |
| | ↓ Not allocated ↓ |

Address markers (right edge): 0xF5007FE0, 0xF5007FD0, 0xF5007FC8, 0xF5007F80, 0xF5007F7C, 0xF5007B78.

Table 4.1: System RAM memory allocations following Node-Boot from ROM.

**Tightly Coupled Memories** Each processor checks its local tightly-coupled instruction and data memories (ITCM and DTCM) exhaustively with multiple test-patterns across all memory locations. If a TCM error is detected then the core is disabled as TCM memories are considered critical to Node- and System-Boot stages and the error is logged as per tables 4.1 and 4.2. If there are no faults then Node-Boot proceeds.

**POST Self-Test for Processor Blocks** The processor block POST tests begin by exercising the non-volatile registers for each processor's Communications, DMA, Timer, and Vector Interrupt Controllers in turn. If a failure is detected in any of these components the processor is not functional; the reason code (table 4.2) is written to the failure log for that processor (table 4.1), the processor has its interrupts disabled, and is put into a low-power sleep mode (effectively shut down).

## 4.1.2 Monitor Processor Arbitration

It is necessary to select a Monitor Processor (MP) which handles the chip-level management functions for the node, including the progression from Node-Boot into subsequent stages of system operation. If the reset cause is a power-up then the list of historic Monitor Processors (table 4.1) is reset, so that all functional processors have a chance to become elected. If the reset reason is *not* a power-up, this may be due to an

## 4.1. NODE-BOOT

| Peripheral | Method | Executor | Failure response | Err Code |
|---|---|---|---|---|
| *At Power-on* | | | | |
| ITCM | RAM test | All | Shut down core | 0x4 |
| DTCM | RAM test | All | Shut down core | 0x2 |
| Comms controller | Register test | All | Shut down core | 0x0 |
| DMA controller | Register test | All | Shut down core | 0x1 |
| Timer | Register test | All | Shut down core | 0x9 |
| VIC | Register test | All | Shut down core | 0xA |
| *After Monitor Election* | | | | |
| Previous Monitor | Check Register | All | Shut down core | 0xC |
| System RAM | RAM test | Monitor | Shut down MP | 0x7 |
| Router | Register test | Monitor | Shut down MP | 0x6 |
| Watchdog | Register test | Monitor | Shut down MP | 0xB |
| PL340 | Register test | Monitor | Record, continue | (see: |
| SDRAM | RAM test | Monitor | Record, continue | (fig.4.6 |

Table 4.2: Ordered list of power-on self-tests performed during Node-Boot. An error code is written to the processor's failure log in the indicated bit position if detected.

error with the Monitor Processor. To prevent this failure recurring, if a processor has already been a Monitor, it will shut itself down and become ineligible for election to become the new Monitor Processor. The chip is now ready to elect a Monitor Processor which is achieved by *mutex* hardware in the System Controller. The first processor reading back the register following a reset event will be installed as Monitor Processor and marks itself as having undertaken the Monitor rôle in the bit-wise 'Monitor History' System RAM register (table 4.1). Subsequent processors will become Application Processors, and wait for the Monitor Processor to complete node-level hardware initialisation, and signal Application Processors to progress.

### 4.1.3 Chip-Level POST and Initialisation

This and subsequent sections are illustrated by the 2nd flowchart, (fig. 4.5). The Monitor Processor continues testing the critical chip-level System RAM, router and watchdog controller, with any failures marked in the failure log (tables 4.1 and 4.2) and the Monitor Processor shutting itself down. This effectively disables the chip, the Application Processors are signalled to continue into sleep mode. In subsequent stages of boot a neighbouring chip may read the cause-code, and it may be possible to 'nurse' the node back to some level of functionality, particularly to enable its router.

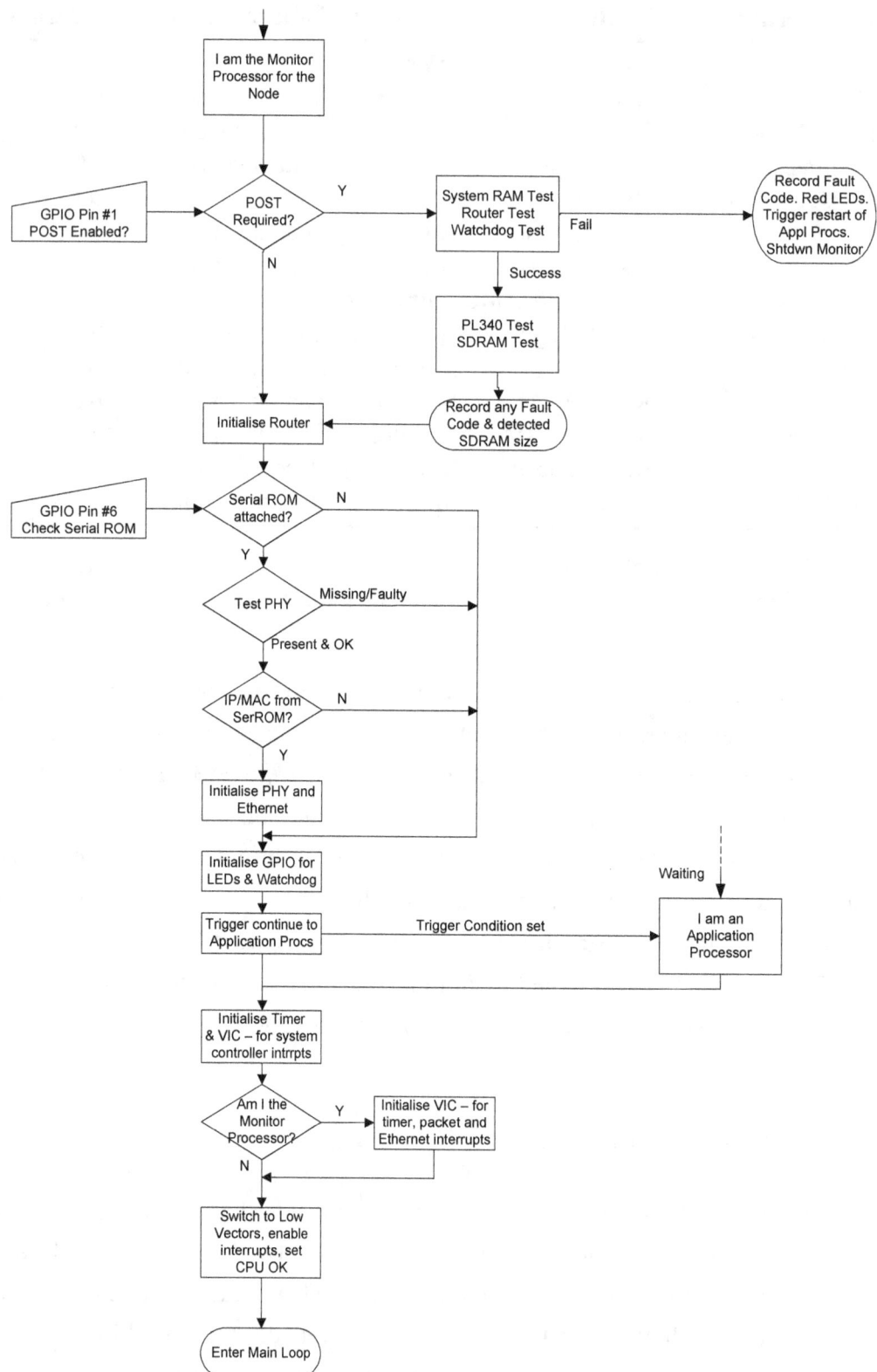

Figure 4.5: Node-Boot flow of processors immediately following Monitor Processor allocation, until entering the listening state in the main loop. (Follows from fig. 4.4).

## 4.1. NODE-BOOT

**SDRAM Testing** This POST is designed to operate slightly differently from the other tests, as any failure detected in the SDRAM or its supporting hardware does not result in the shut down of the Monitor Processor; consequently the chip remains active. Failure information is logged in the 'SDRAM Information' registers in System RAM. The testing begins with checking and initialising the RAM controller, a failure of which results in further SDRAM testing being aborted. If the RAM controller locks to functional memory, tests are performed on the actual RAM to determine its size, and any errors when reading / writing samples of memory are also noted. Performing a full memory test at each boot is too time-consuming, with a comprehensive test (as used on the TCMs) taking 138 minutes (15.5 kB / s). It should be noted that all SDRAM memory die used in manufacture are 'known-good' components.

The sample SDRAM tests are performed at word positions $2^n$, with any fault being noted in the error log at bit position 1-29, representing word $2^{bitpos}$ in SDRAM. The discovered SDRAM size (in bytes) is stored in the SDRAM size register – see figure 4.6. A thorough RAM test is performed on both the first and last 16 Bytes of memory. This is a small sample of a full memory test using a greater combination of data and addressing lines. Any errors found are noted per figure 4.6, and recorded in the SDRAM information fields (table 4.1).

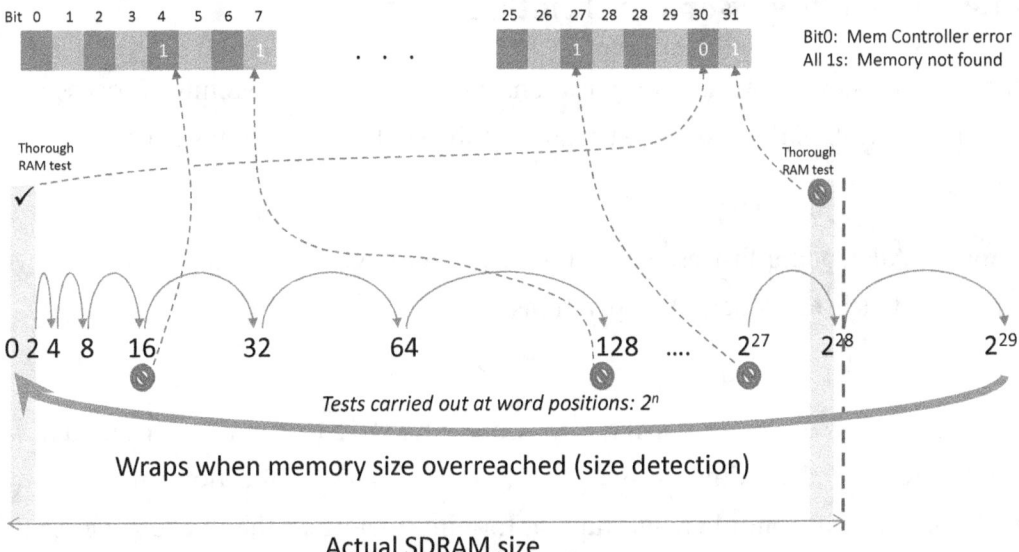

Figure 4.6: SDRAM testing in the ROM. Size is determined by writing at increasing word powers of 2 until the wraparound is determined. More thorough tests are carried out on blocks at the top and bottom of detected memory.

**Router** At power-up SpiNNaker chips are not numbered, therefore the point-to-point routing tables are set up to drop all traffic rather than being left indeterminate. For the Multicast routing tables all 1024 entries are cleared so they do not match any packet, the outcome being a fail-safe configuration of only default routing, where received packets traverse the router and egress the chip on the opposite external link to which they arrived.

**Ethernet** All SpiNNaker chips have Ethernet controller hardware on-board, but only when this is connected to an external PHY chip and Serial ROM is connectivity possible. The unique addressing information is retrieved from the Serial ROM chip (section 4.1.1). In the case of the DHCP Node-Boot image, the IP network address and settings are learned in response to requests made by the image.

**Watchdog** The Monitor Processor sets up the chip watchdog timer, and if it is neglected for 2.5 seconds the chip will reset with a watchdog cause-code. In normal Node-Boot operation this will not happen as the watchdog is explicitly refreshed periodically.

### 4.1.4 Final Processor-Level Initialisation

Once the chip-level tests and initialisations are completed the Monitor Processor signals all Application Processors to progress to the next initialisation stages:

**Timers** All Monitor Processors set up their timer to give a tick approximately every 1 ms, which is used by periodic operations.

**VIC** Each processor's Vector Interrupt Controller (VIC) is set up to respond to particular interrupts depending on the processor rôle. The Application Processors rely solely on System Controller interrupts informing them that they have a message to be processed (in the mailbox, see table 4.1), whereas the Monitor Processors receive interrupts when packets are received from the inter-chip links, by Ethernet frames arriving, and the timer interrupts as mentioned above. When a processor is not dealing with interrupts it enters a low-power wait-for-interrupt 'sleep' state.

### 4.1.5 The Main Loop

Upon completion of self-testing and initialisation, all processors enter the main loop of the program, described in the pseudo-code listing below:

```
while(true) {
    sleep_and_wait_for_interrupt();
    if(i_am_monitor) refresh_watchdog_counter();
}
```

The main-loop is the 'kernel' of the event-driven operation of Node-Boot: processors are placed into a low-power wait-for-interrupt state and wake up only in response to their specific interrupts. On returning from an interrupt, processors are immediately put back into wait-for-interrupt state, with the exception of the Monitor Processor which periodically refreshes the watchdog counter. All processors are now in 'listening' mode in Node-Boot, which ends when a complete System-Boot image has been received by the chip via a flood-fill, and that image begins its execution.

## 4.2 Loading the System-Boot Image

For the SpiNNaker machine to progress beyond Node-Boot new operating code must be pushed to it. In the design process for the chip it was decided to keep the Node-Boot image in ROM simple (and therefore more likely to be reliable), and in operation for it to remain in a passive state awaiting its instructions. All nodes at power on / reset have elected a Monitor Processor which is listening on the node's six external inter-chip links (and where provisioned on the Ethernet connection too) for their 'System-Boot' image, whose functions include preparing the system to load application data and code, performing any further checks / diagnostics required, and facilitating the numbering of the homogeneous SpiNNaker nodes.

The process begins with the responsible Host system seeding the transmission of the System-Boot image, to one or more of the Ethernet linked 'root' chips. The code is assembled and checked by the Monitor Processor on each receiving chip, and the validated code is executed by this core. The first task of the System-Boot image as it starts is to self-propagate its own image to the node's immediate neighbours over the chip-to-chip links (a process known as flood-fill).

## 4.2.1 Ethernet Flood-Fill

To facilitate the automatically discovery of SpiNNaker nodes by the Host, each chip having an active PHY transmits out a broadcast 'Hello' message on the local Ethernet every 4 seconds triggered by the timer interrupt. These broadcast destinations ensure that the 'Hello' messages get to the (unknown) Host, and are only transmitted once every 4 seconds to ensure they are not burdensome to the local LAN. An example of a SpiNNaker 'Hello' message can be seen in figure 4.7.

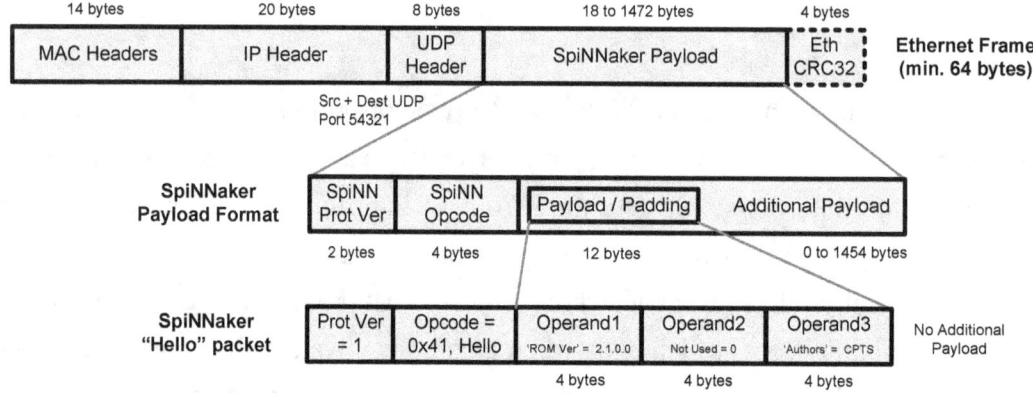

Figure 4.7: The basic SpiNNaker Ethernet framing / packet format (used by Node-Boot). An example of the 'Hello' SpiNNaker packet is illustrated, broadcast by each Ethernet-attached node every 4 seconds.

This same simple SpiNNaker packet format is used by the Host to transmit the System-Boot image across to the machine (fig. 4.8) using 3 different packet formats:

**Flood-Fill Start** When a SpiNNaker chip receives the Start message it readies itself for receipt of an image of up to 32 kB split up into the number of blocks indicated in the start message (a range of 1–256). The image is assembled in the top half of the Monitor Processor's DTCM (fig. 4.8) as it is large enough to collate a full ITCM image. A receive array is initialised empty with an entry for each Block ID expected.

**Flood-Fill Block** Data is now transmitted block by block by the Host to the SpiNNaker system. Block IDs are numbered beginning 0, and the block size in words is indicated (a range of 1-256 words). Typically 32 blocks of 256 words (1 kB) are used, as this has the lowest overhead for production use. If a transmitted block is successfully received then the data is copied to the appropriate position in the DTCM image assembly area, and the receive array is updated to indicate that the block is in place.

## 4.2. LOADING THE SYSTEM-BOOT IMAGE

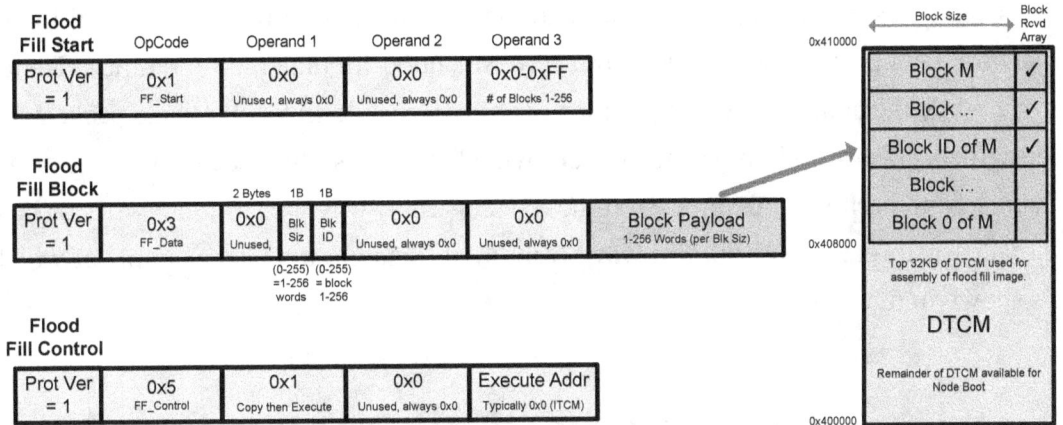

Figure 4.8: The three packet formats used by the Host pushing a System-Boot image to a SpiNNaker node, and the assembly of the data blocks into the image in DTCM.

**Flood-Fill End** Following transmission of the Start and all Block packets, the Host transmits the Control packet. When a SpiNNaker chip receives this it firstly validates that all the blocks in the receive array are populated. If they are not (for example a frame has been lost), then it continues to listen for the missing Flood-Fill Blocks. Typically however, all blocks are in place at the first attempt, and the system copies the assembled image over to ITCM – and branches to the indicated start address.

### 4.2.2 Inter-Chip Flood-Fill

Sending the image from the Host to Ethernet-attached 'root' chip(s) is the first stage of the flood-fill process. The second is that the System-Boot image should flood-fill itself out to its immediate neighbours. This is done on a link-by-link basis using Nearest Neighbour (NN) packets (other packet types are not available at this stage as the machine is not numbered, so the routing tables are not populated). The NN packet type provisions an 8-bit control field, a 32-bit key and an optional 32-bit payload. The payload is nowhere near as large as available in the SpiNNaker packets sent over the Ethernet, therefore an additional level of hierarchy is added to the flood-fill transmissions as each block now requires segmentation. A layered combination of validation methods are also used to check the received image's integrity. As well as simple SpiNNaker packet parity, there is a locally calculated checksum at the packet level, and at the block level the SpiNNaker programmable CRC hardware is employed to generate and validate CRC-32 checksums at transmit and receive ends of the link.

Clearly, as the data propagates around the machine due to the inherent multi-path network connectivity, each node will receive the flood-fill data multiple times. Therefore, while the node is being flood-filled over its inter-chip links, duplicate packets are discarded. Transmissions are not system-wide broadcasts, they occur in waves originating from each node that received the System-Boot software, so there will not be a broadcast storm in flood-fill, just wave-front(s) of transmissions (an example of which is shown in figure 4.9).

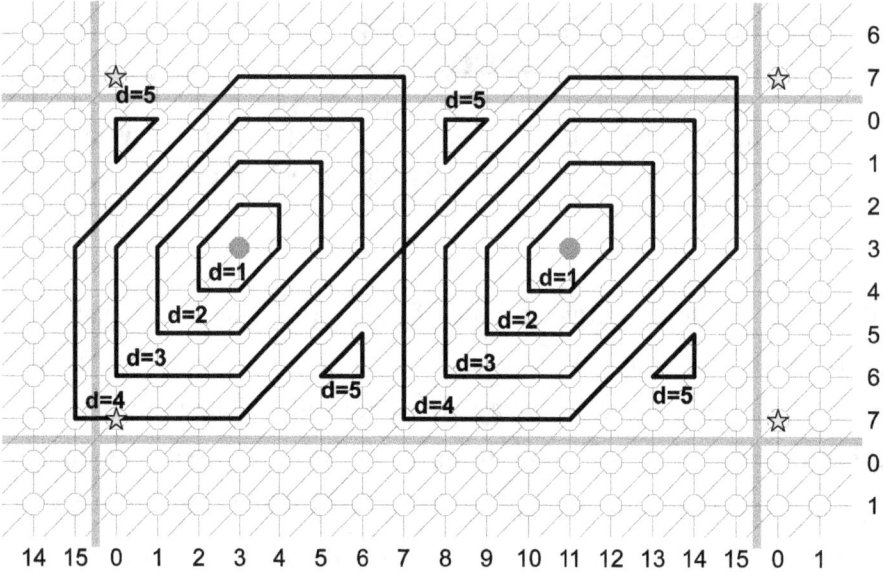

Figure 4.9: 'Waves' of System-Boot flood-filling a tessellated SpiNNaker torus over time. Distance from a seeded root node is noted by $d$.

Due to the small SpiNNaker packet payload, the original model outlined in figure 4.8 for Ethernet is extended to a 5 packet-type model (illustrated in figure 4.16):

**Flood-Fill Start**  This is functionally identical to the Ethernet version

**Flood-Fill Block Start**  The Block start message indicates the block ID, and the size of the block. If this block has not already been completed, a received word array is initialised with an empty entry for each word in that block. Subsequently these data words are listened for. (To take advantage of the CRC hardware the words are collated into the shared assembly block in System RAM (see table 4.1 and figure 4.16)).

**Flood-Fill Block Data**  The block data message indicates both the block ID and the word ID within that block. Words from blocks other than the one being listened for are

## 4.2. LOADING THE SYSTEM-BOOT IMAGE

discarded – only 1 block is populated at a time. The data word itself is carried in the payload, and if it has not already been received is copied into the space used to collate the block. The received word array is updated with words successfully received.

**Flood-Fill Block End**  Firstly a check of the received word array is carried out to ensure that all words are populated. If the list is incomplete then the software waits for a retransmission, however, typically there is a complete set and the data block can now be validated. For the indicated block ID, a CRC-32 is incorporated as the payload of the block end message. At the transmit end this can be generated in software, or via the hardware supported DMA process similarly to the receive checks. At the receive end the CRC check is generated by performing a DMA from the block assembled in System RAM into the appropriate position in the DTCM image, with the CRC option enabled. The received CRC-32 is then compared with the one calculated locally as part of the DMA transfer, and if they match, the entry in the received block table is populated (fig. 4.16). If it fails, then the whole block is discarded as it is impossible to calculate which word is in error (as the checksums are not strong enough for error correction). The system then listens for the next Block Start or Control Message.

**Flood-Fill Control**  This is the same as the Ethernet version.

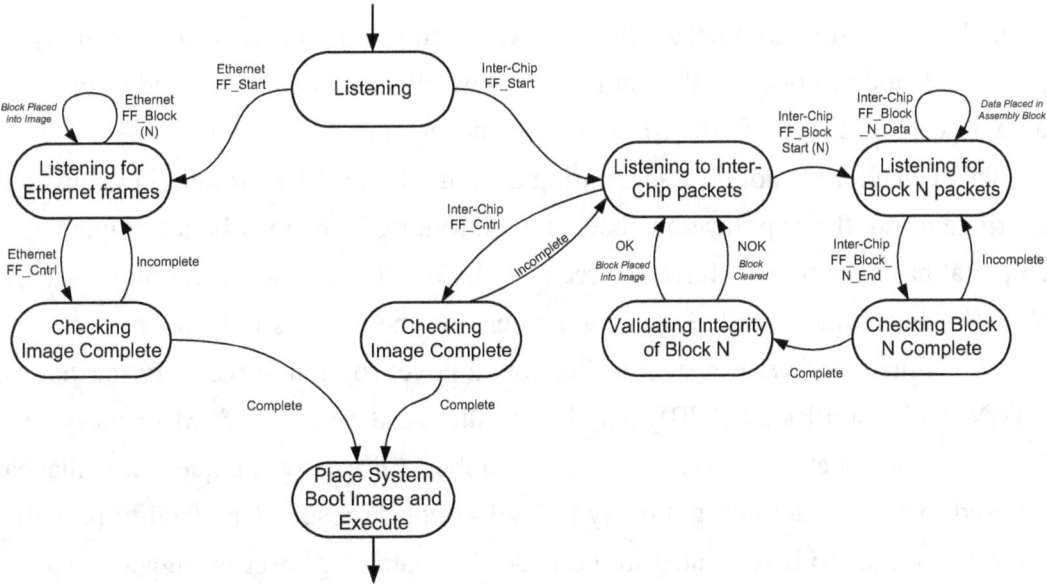

Figure 4.10: The state diagram transitioning from Node-Boot to System-Boot via flood-fill from the Host or a neighbouring chip.

**System-Boot Loading – A state machine**

The state of the SpiNNaker node during the loading of a System-Boot image may be considered as a state machine as detailed in figure 4.10. Note that if a start message is received either on Ethernet or an inter-chip link the state-machine remains in this mode, there is no 'interworking' of the two methods.

**Block size choices and Retransmissions**

All packets on both Ethernet and inter-chip links are transmitted as datagrams, with no guarantee of delivery, so mechanisms are built in to compensate for any loss / errors (await a retransmission of the data). Purely in terms of payload data transmitted, Ethernet is usually twice as efficient as the inter-chip link method with a best case of 94% efficiency, versus the best inter-chip efficiency of 44%.

## 4.3 IVB – ITCM Validation Block – a technique for graceful recovery of Watchdog resets

Should a Monitor Processor within the SpiNNaker machine experience a software fault, it is useful to have enabled the watchdog timer so that the chip is signalled to reset. If a watchdog reset occurs then the whole chip (all 18 cores) would usually return to the Node-Boot code, the routing tables would be reset, and the node would end up quiescent awaiting a System-Boot image once again.

This behaviour is not desirable during a simulation, as the surrounding chips will continue to run their application code, and the routing paths may be interrupted. The chip that has been reset will require recovery before it may once again play an active rôle. However, much or all of the operating environment may still be intact.

To facilitate recovery in such a situation, a novel (optional) recovery mechanism (ITCM Validation Block (IVB)) may be installed at the top of ITCM memory. The IVB is a series of checksums and a 'magic-number' (distinctive unique value) that can be used to circumvent going directly to Node-Boot on reset. This facility permits a recovery routine to be executed in the node if a watchdog reset is triggered and the IVB block is valid, which recovers the core to a 'known-good' state. This known-good state is programmable and could be used to recover the whole ITCM, or run a small routine simply to maintain the routing table entries so the remaining nodes in the machine can continue without disruption. If the IVB facility is initiated, but there has

## 4.4. DHCP NODE-BOOT IMAGE

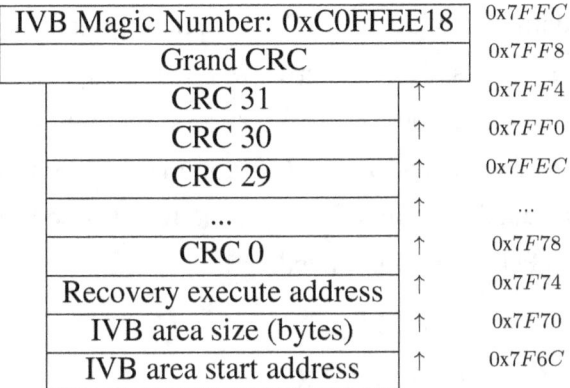

Table 4.3: ITCM validation block structure.

been corruption of the instruction data that the IVB block protects, then the node will enter the Node-Boot routine as normal and await external recovery.

Table 4.3 describes the structure of this optional IVB. This should be populated by the software image which is being setup for recovery, and consists of several fields: Firstly the magic-number is checked – only if this is in place then further checks are performed. For each 1 kB of protected ITCM a 4-byte CRC is stored which is calculated using the programmable CRC on-board the SpiNNaker chip. Furthermore there is a CRC of CRCs (a 'Grand CRC') – this is to ensure that the IVB itself has not been corrupted. The block also contains start and length fields of the contiguous ITCM block to be recovered, and a branch address to be executed if all IVB CRC checks succeed.

## 4.4 Dynamic Host Control Protocol (DHCP) Supporting Node-Boot Image

To aid network manageability of nodes it is necessary to provide a mechanism where a node may learn its Internet Protocol (IP) addressing information, rather than have it statically allocated by the Serial ROM chip. This is useful in cases where a board is mobile and may be attached to many different networks, or where fixed addressing may clash with existing devices on that LAN and cause it an outage or unreliability. In many LAN environments the LAN management team also wish to maintain centralised control of IP address allocations via DHCP so that they can filter devices on and off the network dynamically, so having a DHCP client to cover these sets of cases for the SpiNNaker chip is very useful.

The DHCP standards introduced in [Dro97, Ale97] detail how a host negotiates a lease of a unique IP address from a network dynamically, which remains valid for a specific period of time specified by the server. The solution to this problem for SpiNNaker is an updated variant of the Node-Boot image loaded from Serial ROM (per section 4.1.1), which remains compatible with the ROM based image, but operates a DHCP client once the POST has completed.

### SpiNNaker DHCP implementation

The flow of how a SpiNNaker machine handles the request for a DHCP lease is illustrated in figure 4.11 and begins following the POST and chip initialisation. From the DHCP INIT mode SpiNNaker proceeds to issue a DHCP Discover request, which is a broadcast to learn of any DHCP servers it can communicate with, and passes into the SELECTING state. Assuming a DHCP server responds with a DHCP Offer, the first response is accepted, and the SpiNNaker implementation sends a DHCP Request message to that server to accept its offer, and transitions into the REQUESTING state. The DHCP server, upon hearing SpiNNaker's request (and assuming its initial offer of IP details is still valid) will reply with a DHCP ACK message. If the DHCP ACK details are verified by the SpiNNaker client as the same as in the original offer its state moves to BOUND, and the leased IP details are installed in the Ethernet block (table 4.1) and used by the SpiNNaker system.

The DHCP Node-Boot code has been testing in a variety of DHCP environments to ensure compatibility with different DHCP servers. It has been tested on the University of Manchester LAN, with enterprise-class Cisco routers acting as a DHCP server, and on home-class devices from D-Link and Netgear. In all these cases the DHCP client software function successfully, obtaining IP information initially and successfully renewing DHCP leases based on the lease timers.

### Discovery of the board

When using DHCP for IP allocations, the management of the IP information shifts from being a distributed problem statically coded onto the clients, to a centralised service typically residing in a different management domain. The DHCP allocation of IP addresses is usually automatic from a logical pool of free addresses, and thus the IP address lease assigned to the SpiNNaker board is non-deterministic. If the IP address of the SpiNNaker board is unknown – the management path to the board is undefined.

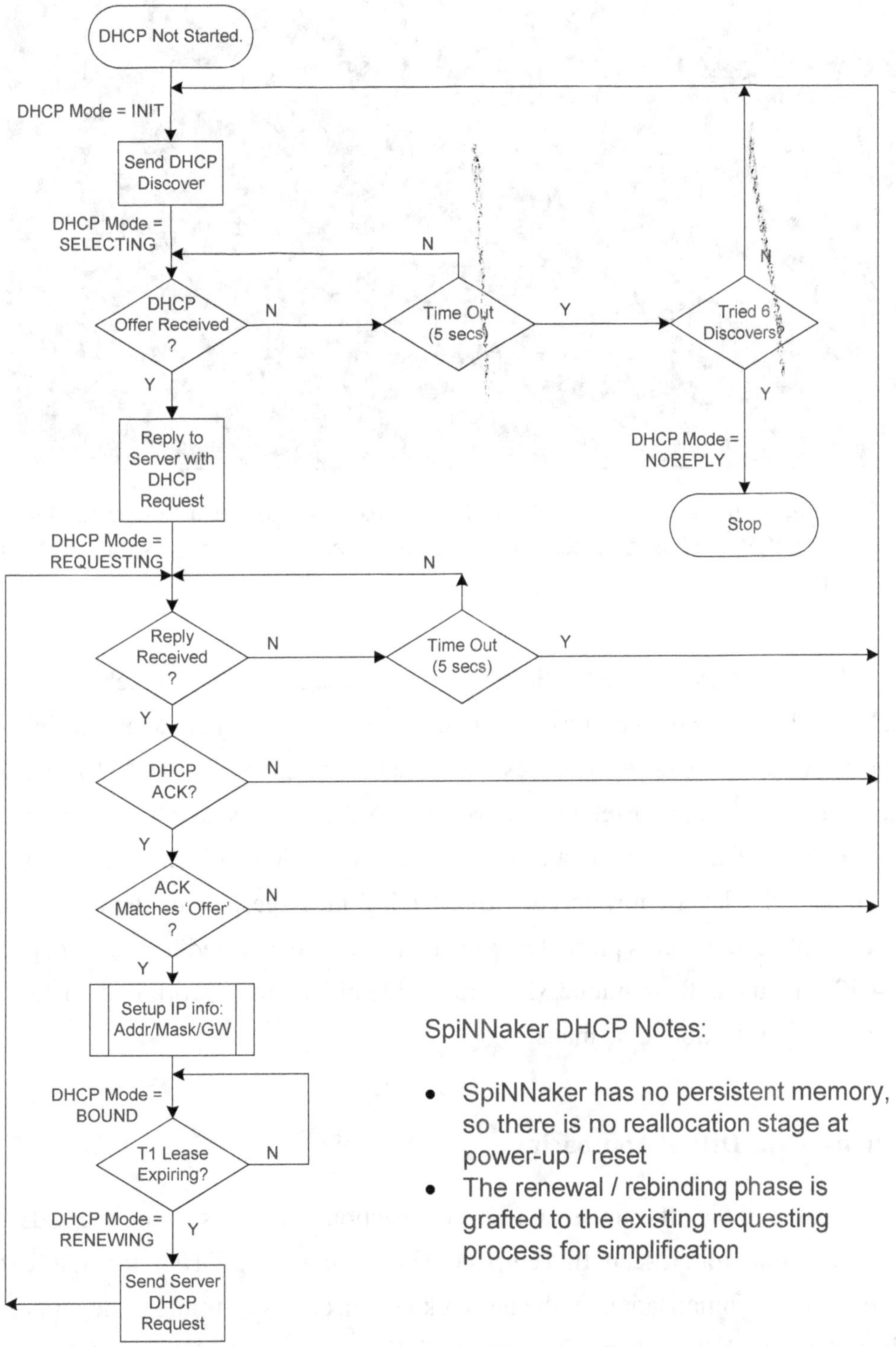

## 4.4. DHCP NODE-BOOT IMAGE

Figure 4.11: DHCP address lease assignment from the SpiNNaker client perspective.

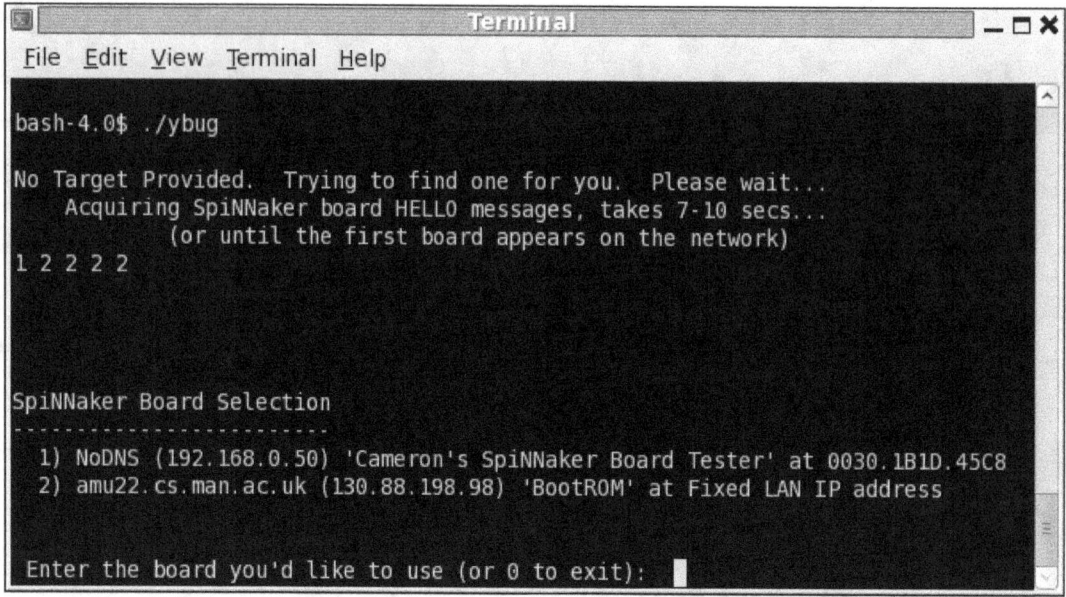

Figure 4.12: Automatic discovery of the SpiNNaker boards on a network. One is allocated via DHCP and has its description text displayed, the other is a standard Node-Boot fixed IP image.

The solution is to extend the 'Hello' discovery message (fig. 4.7) which is sent periodically sent by each Ethernet-attached node at Node-Boot, to include a user-defined 'identification' string of up to 32-bytes recovered from the Serial ROM. Therefore management software may listen to discover SpiNNaker boards attached on the local Ethernet network. The current Host tools have been extended with an auto-detect option for those SpiNNaker chips broadcasting 'Hello' messages. Once the auto-detect period is complete, the user is presented with a menu of the detected boards to connect to (fig. 4.12), including their unique 32 character identifier and network information to allow the correct selection to be made.

**Limitations of the DHCP Approach**

Due to the nature of the off-chip serial ROM connection, using this technique adds 2.5 seconds to the time from reset until a chip's POST is completed. DHCP too adds extra time to the board's initialisation on the network as it takes a short time to negotiate an IP address lease with the server. One further limitation of this solution is that as the SpiNNaker 'Hello' messages are broadcast, they are not routed and remain within the confines of the local network. Therefore for any device to discover that SpiNNaker chip, it must be within this local domain.

## 4.5 Results from Node-Boot Operation

Eighteen core SpiNNaker chips were first delivered in April 2011, and the ROM Node-Boot software is used to operate all test boards. The software appears fully functional and successfully covers the boot sequence from power-on, or reset, through to handover to System-Boot software. The third generation of small (72 core) 4-chip boards (fig. 4.13) were, in June 2012, overtaken in scale by larger (864 core) 48-chip boards, but to perform testing of the flood-fill mechanisms, multiple 4-chip boards have been interlinked in extended configurations.

Figure 4.13: A third generation SpiNNaker 4-chip test board (September 2011).

### 4.5.1 Fault Detection and Isolation

For economic purposes, in the SpiNNaker architecture, the majority of faulty chips will be pressed into service as most faults are likely to affect only a single core. With these chips the faulty core is isolated during power-on self-tests, leaving the remaining 17 processors in operation for Monitor and Application rôles.

The Node-Boot code has been tested with chips that did not pass the full suite of 'manufacturing tests' with equivalent results. The code successfully identified the faults – setting the appropriate bit positions in the core's error code entry (table 4.2), isolating the relevant components and succeeded in booting the rest of the chip's functional cores correctly. However these tests are not 100% comprehensive, see section 9.1.3 for further discussion.

### 4.5.2 Response Time and Data Rate

The Node-Boot image processes three types of traffic on its Ethernet connections: Ping requests, ARP requests, and Flood-Fill messages. Ping and Flood-Fill packets are unicast packets targeted at the MAC address of the node, and ARP traffic is transmitted as an Ethernet broadcast, requiring SpiNNaker to handle the receipt of broadcast packets and resiliently drop non-ARP types. To test all three types of packets, a number of experiments were performed:

**Ping Traffic**

The SpiNNaker board was attached directly to a Host workstation and a volley of ping (Internet Control Message Protocol (ICMP) echo) requests were sent using a range of payload sizes. In response to each ICMP echo request packet, the SpiNNaker chip must copy the payload data, create network headers and calculate a new checksum over the whole packet including payload. Three plots are presented from the data recorded during these ping tests. In figure 4.14a the overall data rate achieved is recorded, in figure 4.14b the number of ping packets handled / s is plotted, and in figure 4.14c the response time is recorded.

The differing network hardware on the Host platforms is characterised by the alternate curves in figure 4.14. Two machines were used for the ICMP testing, each using 32-bit GNU / Linux operating systems. Firstly a Sony Laptop with a 1.4 GHz U9400 Intel Core2 Duo processor and Intel Ethernet hardware, and secondly a generic Desktop PC with a 3.16 GHz E8500 Core2 processor using a Realtek Ethernet chipset. The laptop is not able to sustain either the data rate or packet rate of the desktop machine for smaller packet sizes, with the curves converging with payloads of 512+ bytes. This convergence indicates the true maximum turnaround rate that SpiNNaker is able to achieve for ICMP requests of those sizes. For latency the curves are more widely displaced, indicating that the Sony is able to achieve better latency results, possibly through more capable chip-sets and drivers. Where the graphs are divergent, SpiNNaker is demonstrating its ability to maintain the better of the two performance sets as these tests are sustained over 30 second periods for each testing point, and repeated on multiple occasions.

These results provide a gauge of the performance of the bidirectional network handling code of Node-Boot, with a maximum payload data rate of 0.6 MB / s achieved at around 4,000 packets / s. The maximum packet rate / s of over 69,000 packets is achieved using some of the smallest payload sizes, a gauge of the maximum number of frames that a SpiNNaker can handle / s. For figure 4.14c, the latency rises only very slightly based on the payload length, and appears to not be a limiting factor.

## 4.5. RESULTS FROM NODE-BOOT OPERATION

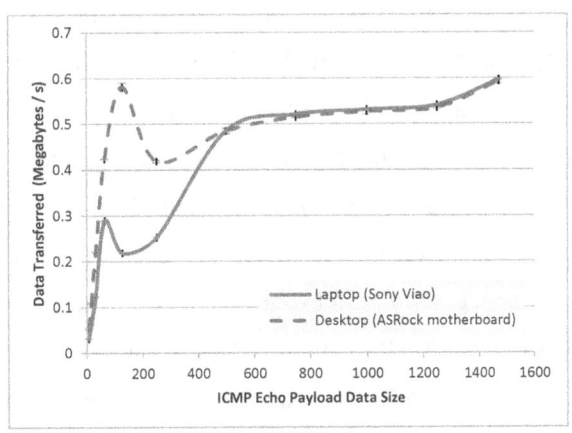

(a) Maximum data rate

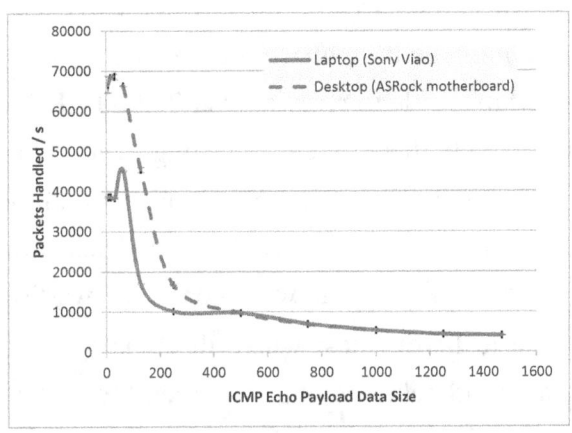

(b) Maximum packet rate

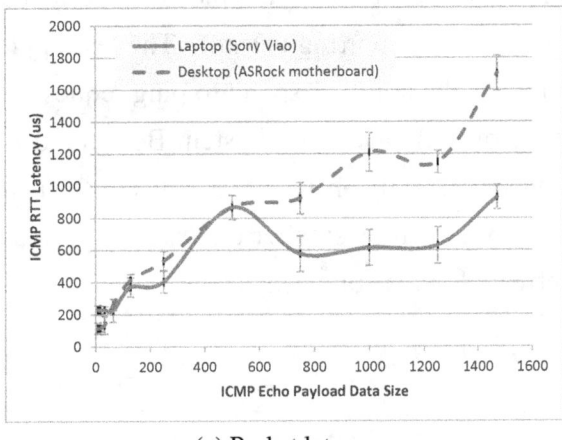

(c) Packet latency

Figure 4.14: ICMP echo request – Node-Boot software performance.

### ARP Traffic

To maintain IP connectivity SpiNNaker is required to respond to Address Resolution Protocol (ARP) requests so that its layer 2 and layer 3 addressing information (MAC and IP addresses) may be mapped. In a typical Node-Boot situation one or more Host systems may be communicating with the SpiNNaker node, and it must be able to handle other broadcast traffic on a LAN correctly. SpiNNaker does not have hardware support for broadcast and multicast filtering, therefore the Node-Boot software must check every multicast / broadcast packet that arrives on its Ethernet. Two experiments were performed to test the Node-Boot handling of broadcast traffic:

Firstly a large number of ARP requests are transmitted from a packet generator and targeted at a SpiNNaker chip. This test is performed directly between the packet-generator and the SpiNNaker chip, as a network switch will usually 'rate-limit' broadcast packets to a port to mitigate the impact of a broadcast storm in a production network. The SpiNNaker chip is able to reply to over 71,000 ARP requests each second, a similar (packet / s) number to that of the ping testing. During the test around 150,000 valid ARP packets / s were sent towards the SpiNNaker chip, more than it can handle, with the excess dropping at the ingress buffer. The SpiNNaker chip remains functional while these tests occur, gracefully coping with the packet overflow.

The second test is a real-world test where the SpiNNaker chip is exposed to a mixed network with a considerable amount of varied broadcast traffic. The University of Manchester's School of Computer Science LAN was used for this test, where a range of 20 to 50 broadcast packets / s are typically experienced over a 24 hour period (mean average over a 10 s reporting period). The test is judged a success if the SpiNNaker board maintains the ability respond to ping requests, (which are reliant on ARP functionality at the outset), and that a System-Boot flood-fill is successful. This test has been performed successfully on 10 subsequent consecutive observed occasions over 24 hour periods, and numerous smaller unobserved periods, particularly during office hours where network load is at its heaviest.

### Flood-Fill Traffic

Tests were performed by sending a full 32 kB image to an Ethernet-attached SpiNNaker chip as quickly as possible. The image is sent as a series of sequential blocks (per section 4.2.1) and is not repeated. The success of this test is measured by the image loading and correctly operating on the target SpiNNaker platform, and on its

## 4.5. RESULTS FROM NODE-BOOT OPERATION

repeatability at this rate of transmission at least 50 times without a fail. As this is a test of the ultimate performance of the Node-Boot code, it is performed on a back-to-back connection with the Host rather than via a contended network switch.

The results of this testing show it is possible to transmit the 34 required packets including a start block, 32 × 1 kB data blocks, and a control block, from a Host system towards a SpiNNaker board within an overall period averaging 3.367 milliseconds (10.4 MB / s when headers and trailers are accounted for). To eliminate buffering as a factor the rate was validated by transmitting the data towards a second workstation running a packet capture program, and recording the elapsed time between first and final packet transmission. It is therefore possible to draw the conclusion that SpiNNaker is able to parse flood-fill data input at 10.4 MB / s on its Ethernet interface, which is approaching line-rate.

The results recorded in these performance tests are the best possible which can be attained from a SpiNNaker system running Node-Boot. In production it is prudent to throttle the rates back (by means of inter-packet interval) to ensure reliable performance in non-optimal configurations. The next results section concentrates on how a large system running Node-Boot responds to the flood-fill of a System-Boot image.

### 4.5.3 Time taken to Flood-Fill a System-Boot Image

Following POST all cores enter a quiescent state, with the Monitor Processor listening for packets received from the inter-chip links (and the Ethernet interface if it is enabled). The Host system injects a software image via Ethernet which is assembled and executed on the Ethernet-attached 'root' chip. This software image propagates itself through the network as a flood-fill (fig. 4.9) using Nearest Neighbour communication – the only available mechanism until nodes are numbered and routing tables populated. The use of self-propagating flood-fill greatly reduces the time taken for System-Boot as this is performed in parallel rather than sequentially. In a rectangular $M \times N$ SpiNNaker network ($M$ being the *smaller* dimension) the diameter ($D$) or hop-count from any one node to its most distant is:

$$D = \lfloor \frac{N}{2} + max(0, \frac{2M - N}{6}) \rfloor \qquad (4.1)$$

The use of well-placed multiple software seeding points in a larger system reduces this distance. In the earlier example of figure 4.9, the one seed diameter is 8 hops, but using two seed nodes reduces the distance to 5 flood-fill hops.

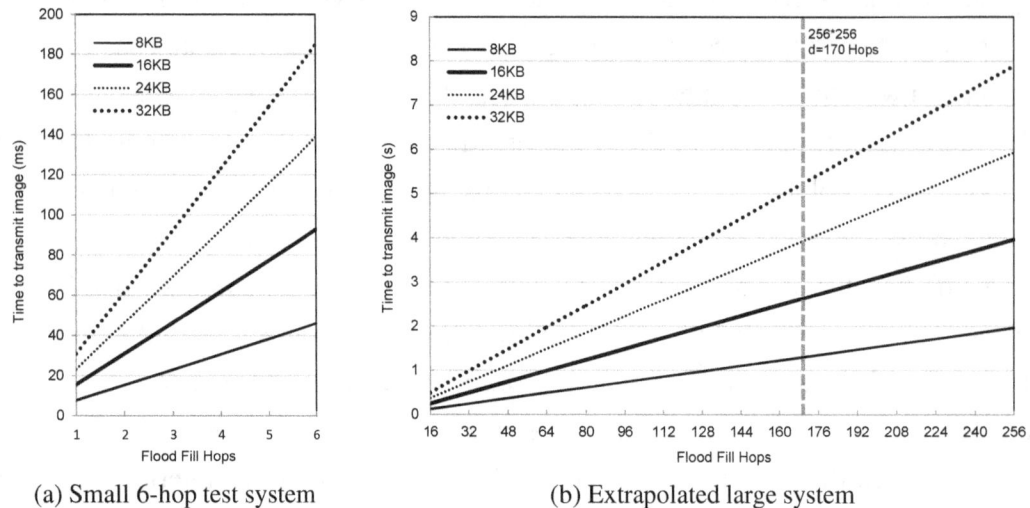

Figure 4.15: Flood-filling the System-Boot image. In a $256 \times 256$ system, maximum distance is 170 hops.

Experiments on SpiNNaker hardware using this self-propagating technique yielded results shown in figure 4.15a, further detailed in [SPF11]. The transmit times prove linear to the number of hops, and the duration of the flood-fill of a System-Boot image for any regular topology can therefore be anticipated (fig. 4.15b).

The topology of SpiNNaker machines is expected to be approximately 'square' (e.g. fig. 3.3). A maximal $2^{16}$ node system therefore (using equation 4.1) has a distance of: $D = 170$ hops from a single code injection point which suggests a maximum System-Boot time of $5.3\ s$ (from fig. 4.15b). Usually the time taken will be smaller, e.g. around $4\ s$ for a 24 kB System-Boot image, and in proportion to $D$ if multiple seed points are used.

If the topology and relative positions of the various injection points are known, this phase can also trigger an algorithm to number the nodes. Once the nodes are numbered, point-to-point routing is possible and the Host can communicate with any node directly. At the numbered stage each individual SpiNNaker node can answer management polls and / or send their status autonomously to the management stations, so that the overall health of the system can be determined, and be used in the mapping stages of the application software and in construction of the multicast routing tables (see section 3.5).

## 4.6 Application Loading

The third and final phase of boot is the Application-Load and run-time stage which follows System-Boot and node numbering. Applications and data are loaded onto each core and memory, but this lacks parallelism, and the time taken scales linearly with the number of nodes and size of the data sets. Exploiting the fact that most cores will be running identical application software, a flood-fill mechanism may be used for this requirement too. Monitor Processors at application run-time will have a different image from the Application Processors, but that too can be replicated across the machine efficiently using flood-fill as has been seen. Larger data sets, however, make up the majority of the loaded data in a SpiNNaker ANN simulation, primarily the synaptic interconnection tables, and they are loaded serially onto a SpiNNaker system. In the future it is planned to make use of the SpiNNaker massively-parallel computational resource to perform the detailed mapping of the system, therefore reducing the size of the data sets transmitted to the machine itself at the outset.

## 4.7 Summary and Contributions

The key contribution from this work is the Node-Boot image itself which is operating on thousands of cores already as part of the 4 and 48-chip deployments. As the Node-Boot code is in silicon read-only memory and cannot be changed, its requirements were that it should be 'right-first-time'. Node-Boot is functioning successfully and efficiently as the mechanism to perform power-on self-tests and to enable transition of all cores on the chips to software operation. It is also logging faults successfully, including on the 17 functional core nodes which will be deployed, permitting the rest of the chip to boot up without impact.

Particular points of novelty are:

- The implementation of the ITCM Validation Block (IVB), a mechanism permitting users to subvert the usual reset procedure after a software fault triggers a watchdog reset to minimise disruption
- Exploiting the hardware CRC mechanisms to checksum flood-fill messages
- The method of SDRAM sizing / error detection
- The auto-discovery mechanism for SpiNNaker nodes (particularly with DHCP)
- The immutable data copy / execution routines enabling full ITCM and DTCM images to be used in all stages of boot

From a performance perspective testing has identified that it should be possible to pass and execute a full 32 kB System-Boot image across the full diameter of a full $256 \times 256$ system in just over 5 seconds. This time turns out to be trivial in relation to the time taken for generating the routing and data structures from simulation, and for loading this information to the actual system in latter stages of boot.

From a management perspective, the system successfully bootstraps the machine and logs information regarding the hardware health of the system in the correct locations. This information is available even if the node is shut down due to a node-level critical fault, and may be used to nurse some level of functionality from the node in later stages of boot should this be beneficial for overall system operation.

## 4.7. SUMMARY AND CONTRIBUTIONS

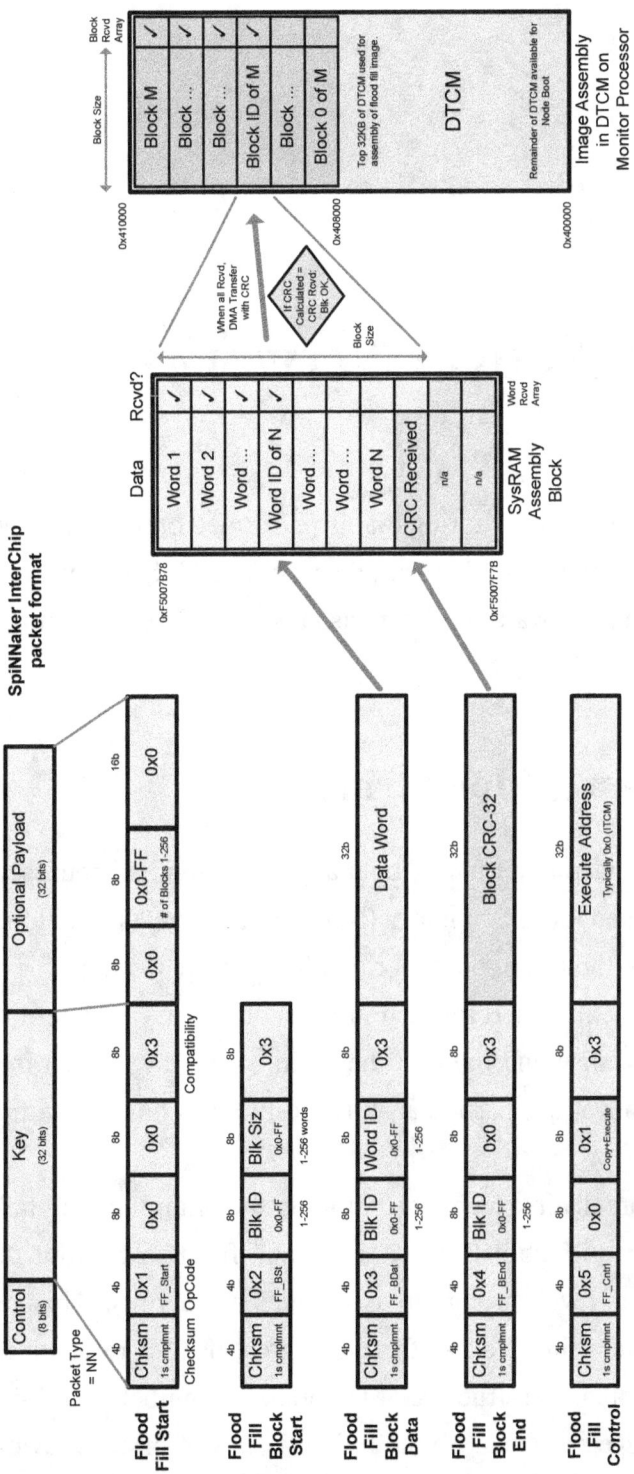

Figure 4.16: The five packet formats used by the chip-chip flood-filling of a System-Boot image, detailing the staged validation and assembly of the data into the image.

# Chapter 5

# Imaging Neural Networks

This chapter introduces monitoring that is performed on neural networks. Similarly to chapter 2 it begins from a biological perspective, and then covers the artificial neural network space and software used for visualising ANNs, both for in-flight and post-simulation analysis.

## 5.1 'Wetware' Monitoring

'Wetware' is a vernacular term used by many computational neuroscientists to describe the in-vivo brain and its constituent cells in animal subjects (including humans). Study of the anatomy of the brain, post-mortem, has been performed for hundreds of years, particularly for teaching, and for the purposes of autopsy to determine cause of death. The ability to visualise and monitor the brain whilst the subject is still alive is a relatively recent innovation, particularly as the brain is housed within the skull for its protection.

Today there are diverse techniques used in the brain imaging field, from structural imaging through to functional monitoring; and from single-unit recordings that can be made at the cell-level through to assemblies of neurons. This section looks at the imaging of the brain 'in vivo', as there is a great deal of research into understanding the function as well as the structure of the brain. A number of non-invasive techniques have been developed to visualise the living tissues of brain; previously the only way to observe the organ in a living patient or animal was through a craniotomy where a section of the skull is removed to expose the brain tissue beneath.

## 5.1. 'WETWARE' MONITORING

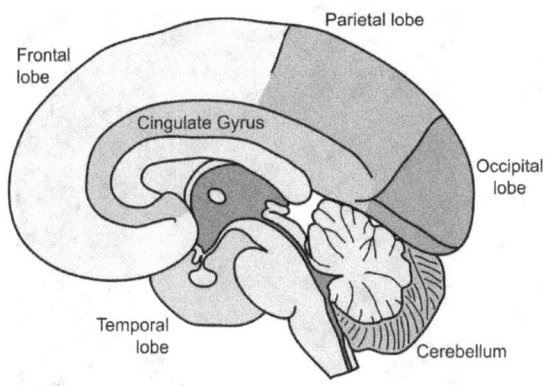
(a) Major regions of the human brain (src: NEU-ROtiker on Wikimedia)

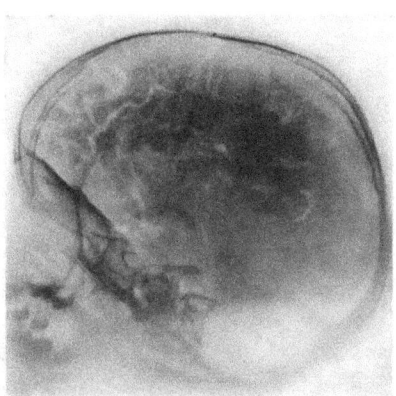
(b) X-ray of brain using pneumoencephalography (src: Wikimedia)

Figure 5.1: Structural representations of the human brain.

There are two main reasons for imaging the brain of a living subject: to explore the structure of the brain for medical reasons – searching for anomalous formations, and secondly for research purposes to learn more about the functioning of this complex computational organ. A greater understanding of the brain as a whole may lead to improvements in the treatment of cognitive disabilities and with degenerative diseases or trauma. Exploiting the techniques employed within the brain would also greatly benefit computer science as the brain undertakes massively-parallel computation with very low power consumption – mirroring exactly two significant problems facing general purpose computing today.

An illustration of the major structures of the human brain can be found in figure 5.1a. Imaging of the brain can be performed structurally or functionally – the former is analogous to a snapshot photograph of the brain, the latter is able to detect brain activity levels – how the brain is actually functioning at the time of the scan. In functional imaging multiple images are typically taken to form a sequence of changing activity levels within the brain, and is used with reference to the applied stimulus to discern functional areas. Newer techniques apply tomography to imaging which effectively allows analysis of arbitrary tissue 'slices'.

The following sections briefly cover the main brain imaging techniques – beginning with the anatomy and ending with function – providing a background to their use, their strengths and weaknesses and examples of their output.

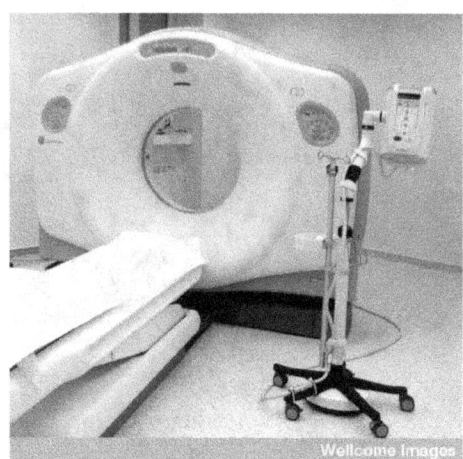
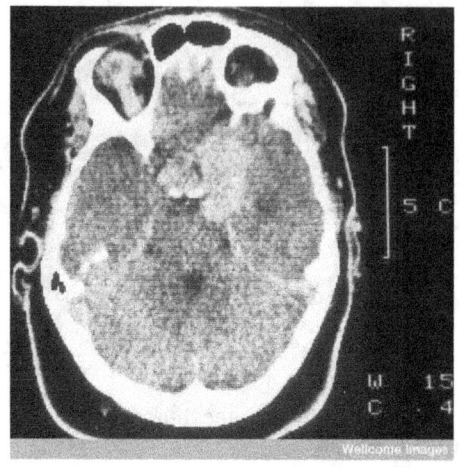

(a) A LightSpeed CT scanner installation

(b) A CT scan detailing contrast in the brain structure

Figure 5.2: Example of machine and data output from Computed Tomography (CT) equipment. (src: Wellcome Images).

## 5.1.1 In-Vivo Structural Imaging

### X-ray Based Anatomical Techniques

**Plain X-ray** Plain X-rays [Ron96], were the first method used for brain imaging, but have mainly been superseded by newer improved techniques in neuro-imaging. X-rays are able to pick out some brain anomalies, but due to the fairly uniform density of the brain [BHK+99], and its soft tissue constitution – its structure is not readily discernible on the output plate or detector.

**Pneumoencephalography** In PEG [Dan18], the inter-cranial fluids in the ventricles of the brain are drained via spinal tap, and replaced with gases to generate better contrast in the X-ray images. An example image using this technique can be found in figure 5.1b, but even this technique exhibits poor contrast. PEG itself is a significant procedure, requiring an extended recovery period for the patient, and has since been superseded by many newer, less intrusive techniques described below.

**Computed (Axial) Tomography** Computed (Axial) Tomography [SAL06], (more commonly known as CAT or CT), is an imaging technique that uses computer technology to control a targeted X-ray source and sensors which are applied radially around the subject (fig. 5.2a), and to combine the series of plain X-ray data sets. The images created are therefore tomographic 'slices' of the subject, and can furthermore be

## 5.1. 'WETWARE' MONITORING

presented in any orientation desired, targeted at the area of investigation [SAL06]. In a head CT (fig. 5.2b) the sensitivity of this technique far surpasses X-ray, with differing tissue densities represented clearly, so white matter can be distinguished from grey matter, tumour from healthy tissue and in trauma haemorrhage can be easily and quickly identified. As computing technology has improved the time taken to create images from CT scans has reduced such that near real-time visualisation of the data may occur as the scan progresses.

The main issue with all X-ray based techniques is that they actively expose the patient to radiation and, cumulatively, this may have an adverse effect on subjects increasing their lifetime chance of developing cancer due to their additional exposure to the ionising radiation used [SBLM$^+$09].

**Magnetic Resonance Imaging Techniques**

Magnetic Resonance Imaging (MRI) [Lau73] dispenses with X-ray and radioactive techniques and concentrates on how sub-atomic particles react to being placed in a strong electro-magnetic (EM) field. In such an environment the protons in hydrogen atoms orient themselves uniformly to the field (the proportion that do so depends on the strength of the EM field). This field is then disrupted selectively by a radio frequency (RF) burst causing hydrogen protons to tip to a different alignment and enter a higher-energy state. The RF energy is turned off, and the alignment of the hydrogen protons returns to the direction of the static magnetic field, emitting RF energy as they relax back to their magnetised state. The rate at which photons are released during the process is dependent on the tissue type, and is detected by the scanner and measured. Different tissue types have different hydrogen compositions, so the amount and rate of emitted RF alters depending on its type. The positional information is retrieved by addition of gradient magnets in 3 dimensions which, when selectively turned on, can modify the magnetic field gradient along the associated axis. The specific frequency required to tip the nuclei now changes spatially over the length of the axis to produce the 'slice'. Analysis on the phase and frequency of returned RF information can then isolate the position in the remaining 2 axes, and form the image (eg. fig. 5.3).

**Diffusion Tensor Imaging** This variation of MRI scanning examines the connectivity of parts of the brain by observing the motion of water molecules through the tissue. By examining water molecule diffusion over time at a particular position, geometric structures (and nerve fibre flows) may be inferred. For example water diffuses

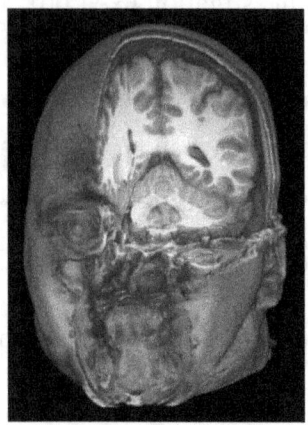 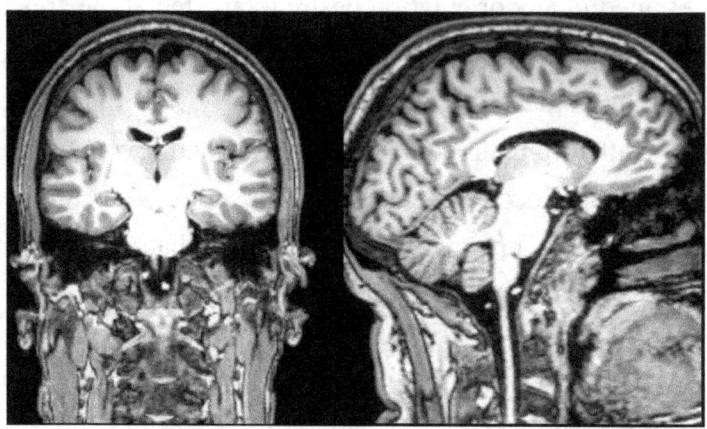

(a) Volumetric representation of MRI structural data

(b) Sagittal and axial views. Cortical (grey) and connectivity (white) matter can clearly be discerned

Figure 5.3: Images of the author from the 3T scanner at Salford Royal Hospital, participating in a brain imaging study. Results of this study were published as [WLK+11].

more readily along an axon than perpendicular to it due to the constraints of the fibre's sheath. DTI studies are carried out in 3 dimensions as facilitated by the MRI imaging, and thus enable a detailed map of the connectivity structures of the brain to be built.

## 5.1.2 In-Vivo Functional Activity Imaging

**Positron Emission Tomography**

Positron Emission Tomography (PET) [TPPHM75] is a technique that uses a radioactive tracer agent, which has a short half-life, injected into the subject's bloodstream. Areas in the brain which are active use comparatively more oxygen and glucose than others, and if the tracer agent is attached to the glucose then proportionally more of it will be found in the most active portions of the brain. The radioactive agent emits positrons which when annihilated with a nearby electron emits two photons (gamma radiation) which disperse at 180° from one another and the time of arrival difference at the detectors may be used to identify the positional source. The scanner may now 'light up' this particular area of the brain image using detection incidences to determine activity levels in that particular brain area (fig. 5.4a). A PET scanner is formed of a ring of sensors physically resembling the CAT scanner (fig. 5.2a), and is typically used in conjunction with CT or MRI scans to co-register both structural and functional data onto the same image (fig. 5.4b). This practice of augmenting a structural scan with overlying functional data is commonly known as imaging fusion or co-registration.

## 5.1. 'WETWARE' MONITORING

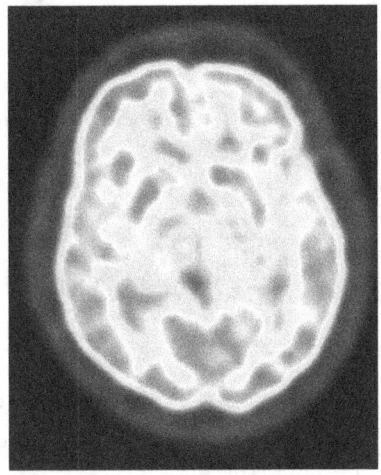
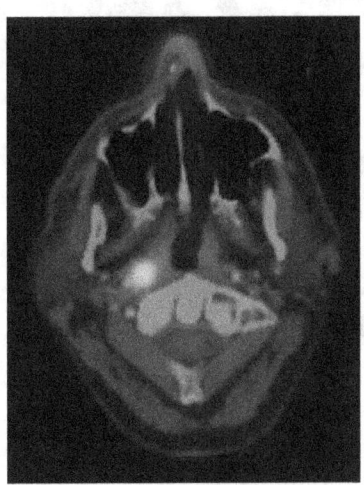

(a) An image from a PET scanner showing radioactivity detected indicating increased blood flow and hence activity

(b) An additional PET image overlaid on CT anatomical detail

Figure 5.4: Example of raw and augmented use of PET data for brain activity analysis (src: Wikimedia).

**Single Photon Emission Computer Tomography**

Single Photon Emission Computer Tomography (SPECT) [JCSM79] is a technique which uses a gamma emitting radionuclide in the bloodstream of the patient which is bonded to a secondary chemical that will be prevalent in the area of interest for the scan. Gamma detectors rotate around the patient over the period of the scan and from the 2D measurements of the gamma radiation detected, a 3D view may be constructed. SPECT is less spatially accurate in its detection than PET, but uses more easily obtained radionuclides

**Functional Magnetic Resonance Imaging**

Functional Magnetic Resonance Imaging (fMRI) [BKM$^+$91] uses rapid (sub-200ms) MRI scans to image brain functionality in semi real-time. fMRI operates by detecting the slight differences in blood oxygenation within the brain and utilises the 3D positional resolution available in MRI [dLVA$^+$98] to pinpoint the functional activation centres (fig. 5.5). The technique generally used is Blood Oxygen Level Dependent (BOLD) [OLKT90], but it should be noted that BOLD does not measure the neural activity directly, but a secondary effect of the haemodynamic system replenishing de-oxygenated blood with oxygen-rich blood following periods of increased activity.

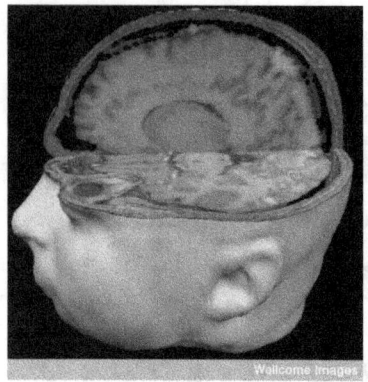 

Figure 5.5: A cutaway functional MRI scan, highlighting visual cortex activation (src: Wellcome Images).

Figure 5.6: A SQUID skullcap worn by subjects in MEG studies, (src: [VMM+04]).

There is therefore a 'lag' with this indirect measure when looking at neuronal activity relative to some other techniques. fMRI is the most common technique used by researchers studying brain activation and function, as it can image activity in three dimension of the brain, and fuse this with a structural scan taken in the same scanner at the beginning of the study; also, of course, it does not involve ionising radiation.

**Magnetoencephalography**

Magnetoencephalography (MEG) [Ham92] is a technique where magnetic fields generated inside the brain are measured externally via devices known as Superconducting QUantum Interference Devices (SQUIDs) which are usually embedded in skull caps / helmets worn by the subject (fig. 5.6). Measurements are taken within a heavily shielded room, as the magnetic signatures generated by the electrical current flows between neurons are very weak. Only assemblies of thousands of active neurons at a time can actually be detected, even in the shielded room. One of the advantages of this technique over fMRI is that MEG is the observation of the direct effect of activation, rather than a secondary effect within the brain, and thus its temporal resolution is consequently much greater.

**Electroencephalography**

Although not a technique which readily lends itself to brain imaging, the non-invasive electroencephalography (EEG) [SG98] technique makes uses of electrodes placed on the scalp (fig. 5.7a) to measure electrical activity of synchronised neural network activity in the cortex of the brain. This technique differs from MEG due to its monitoring of

## 5.1. 'WETWARE' MONITORING

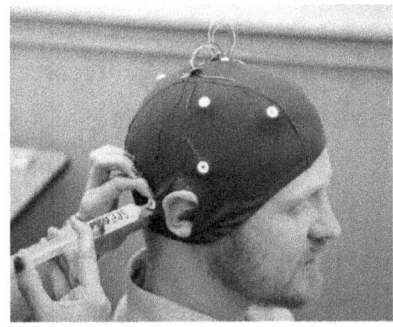
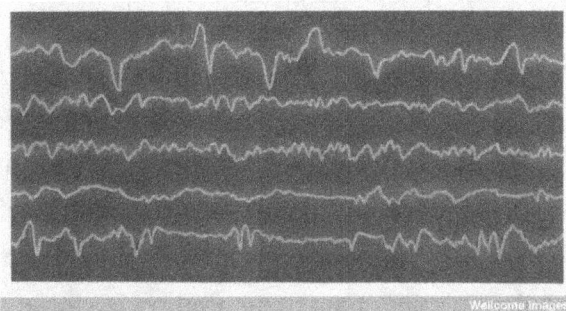

(a) Application of a 10 electrode skullcap for EEG experimentation

(b) An example trace from an EEG experiment (src: Wellcome Images)

Figure 5.7: Electroencephalography (EEG).

electrical rather than magnetic artefacts and may suffer distortion due to the electrical characteristics of the scalp. The typical output from an EEG trace is provided as an example in figure 5.7b. EEG is often used for Brain Computer Interfacing (BCI) where brain activity may be read passively from the subject using relatively simple sensing equipment (fig. 5.7a), and used as input for a computing environment.

**Near Infra-Red Spectroscopy**

NIRS is an optical method that may be used for monitoring the cerebral haemodynamic system [Job77]. The wavelength of near infra-red light is able to pass relatively unimpeded through the scalp and skull where it is either absorbed or scattered by the underlying tissue. The absorption spectra of oxygenated and de-oxygenated haemoglobin differs and hence the returning scatter may be used to record the regions of the cortex which are active with good temporal resolution. This measurement can be repeated across the scalp and thus optical recordings can be made of the activation of functional areas of the brain cortex, the spatial accuracy of which can be improved by the use of multiple optical detectors. In smaller brains (infants or animals), where light may pass through the entire brain, Optical Tomography (OT) may be performed by this technique, although this is not possible in the adult human brain. The technique is typically carried out using laser light sources and the heating function of the absorption is a limiting factor in the source powers used for monitoring.

### 5.1.3 Single-Unit Monitoring

Although the techniques described above allow some level of direct and indirect recording of neural activity within the brain, this is recorded without precision either spatially, temporally or in sensitivity. The monitoring therefore takes place at a functional area, or cell assembly level, rather than with individual neuron cell granularity. Single unit monitoring (at a neuron scale) is readily achievable using in-vitro techniques, for example the squid giant axon used by Hodgkin and Huxley for their experimentation [HH52] or in [IHS06]. However in-vivo methods such as Electrocorticography (ECoG) are more problematic as they require physical access to the subject's brain. Even after access is provided issues pervade in-vivo sensor location – as neuronal structures are small with cell bodies in the micrometre range – and placement cannot be made with this precision. Results recorded may be the result of several proximate cells (multi-cellular recording) rather than a single neuron. The brain also proves to be mobile due to the physical movements of the subject (and brain and sensor inertia), the haemodynamics within the skull, and the operation of plasticity. These changes are large in comparison with the placement precision of the individual micro-electrodes or microelectrode arrays (fig. 5.8), and their attachment to the skull or cortical tissue [MAS11]. The sensor environment is also harsh, there is biological resistance to foreign objects embedded in the body, and over time sensors may become less effective as they react with the surroundings. These issues have led to attempts to create movable single unit sensors [MAS11] to maximise the quality of long-term implanted micro-electrodes. In-vivo single neuron sensing remains an active area of neuroscience research, with high-profile results such as the 'Marilyn Monroe neuron' [CTM+10] reporting success.

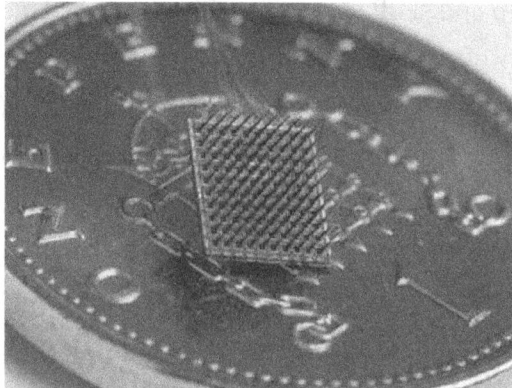

Figure 5.8: A Micro-array of electrodes used for single- / multi-unit monitoring (src: [WC05]).

## 5.1. 'WETWARE' MONITORING

Finally, the implanting of sensors is disruptive to a patient, so this type of single unit monitoring is typically applied in animal subjects, or in humans where there is already some underlying pathological reason for intervention into the skull, for example when implanting brain stimulation devices in cases of Parkinson's disease, chronic pain, depression and Tourette's syndrome [KJOA07].

### 5.1.4 Biological Imaging Software

Much on-line imaging software is supplied to the market as part of a bundle with the imaging equipment itself [BCFG03], and driven by the operator (usually a radiologist). Off-line analysis software is available for various platforms, and particularly GNU / Linux based, in part, on its long-standing support of 64-bit addressing and thus large data sets, and the low acquisition cost (typically nil). Commonly output from structural and functional imaging hardware is in the form of data formats created as part of the Digital Imaging and Communications in Medicine (DICOM) standard [Gib08]. These standards permit software packages to read off-line data to analyse and manipulate it in the software package of choice for the user. Referring once again to the 2011 Frontiers in Neuroinformatics survey paper seen in chapter 2 [HH11], the top 5 software resources used for imaging data as found by the study are:

- SPM [fN12] (Statistical Parametric Mapping) which is a software analysis package for MATLAB [Mat12a]
- FSL [oFOU12] (Oxford Centre for Functional Magnetic Resonance Imaging of the Brain Software Library) visualisation and analysis tools for MRI data
- MRIcroN [Ror12] a set of visualisation tools for displaying volumetric and tomographic slices of medical images. Figure 5.3 was taken from this software
- Freesurfer [fBI12] is a software package focusing on cortical reconstruction from MRI data and functional co-registration upon it
- AFNI [Bet12] (Analysis of Functional NeuroImages) is a C based tool to display functional data from fMRI data sets.

All these tools run on Linux (and Mac), with a few supporting Windows natively (or in virtual machines). They are all free to download and use, and a number are multimodal, in that they can support data from a variety of imaging sources for example MRI, PET and CT. Similarly tools are available for EEG and MEG type data, such as EEGLAB [fCN12] and FieldTrip [CfCNotDIfBB12], both Toolboxes freely available for MATLAB to allow visualisation and analysis of electro-physiological data.

With brain imaging's inherent difficulties of recovering good quality data, and invasive nature to examine the single-cell biology, the case for creation of artificial computational equivalents is strengthened. Simulated models also permit parameters to be obtained through all levels of abstraction in the simulation, in a reliable and noise-free manner, and are the focus of the next section.

## 5.2 Artificial Neural Network Monitoring

Observing an animal provides only the merest hint of what its brain is actually doing, and to understand the activity inside, functional imaging and recording techniques are used as in the previous section. This notion also holds true for computers and electronics running artificial neural networks: machines on desks or in racks appear as 'black-boxes', giving very little away of their internal machinations.

To provide some insight into the operation of artificial neural networks, a level of functional analysis of the simulation and its results are required. As the number of hardware and software tools used for ANN simulations is wide and varied, so too are the techniques used in this area. Some have the analysis of the data taking place offline, with data downloaded from the target platform after the simulation has completed. This is analogous to batch processing in the computing world, but in the neural space may consist of the final weights and biases applied to the components in the network, or in the pattern of action potentials over time. Other analysis methods provide an insight into the performance of the networks whilst the simulations are operating (real-time), perhaps providing a feedback path so that the user may help guide the simulation and direction of learning.

The scientific visualisation [Tuf01, DB91] of neural network data, in whatever form, takes advantage of the brain to discern patterns within the visualised results data, something to which the brain is highly attuned. The representation of the data presented by the visualisation therefore is key to its accessibility, and hence the ability to understand how the neural system operates and reaches its 'decisions'. An issue which pervades visualisation techniques is the high-dimensionality of the data – and how the pertinent information may be represented visually to the user within a 2- or pseudo 3-dimensional space.

This section continues to provide an overview of some of the more commonly used analysis techniques in ANNs, those which operate on the more widespread perceptron based networks and also third generation spiking neural networks.

### 5.2.1 Non-Spiking Neural Networks

The earlier generation of ANNs, as described in chapter 2, are not particularity biologically faithful on a cellular level, but they do encompass some interesting behaviours, particularly in the field of learning of the approximated neural assemblies. Consequently visualisations have been created to represent these behaviours, particularly focusing on how the network 'learns' and reaches its final configuration. It is the network configuration as a distributed whole that encodes the behaviour of the ANN, and it is therefore challenging to represent complex configurations.

**Static Representations**

Craven and Shavlik [CS91] recognised several reasons why the analysis of result data from neural network learning behaviours is important, with visualisation a key technique for achieving this goal. They reasoned that if the network can be understood then the results can be explained with confidence, and that ANNs may be able to pick out previously undiscovered patterns and rules within the input data set. With any learned response care must be taken to maintain the generalisability of a network, and avoid specialising too closely to the training data – over-fitting. However they also recognise reasons why representing a network is difficult – including that the parameters are typically encoded as real numbers, that there are often very high levels of interconnectivity within the system, and that the overall results are a product of the whole distributed network – which may not be trivial in size.

The first eminent method of presenting configuration data from a neural network is the Hinton diagram [HMR87]. The Hinton diagram displays vector information of input / output weights of a unit's connections and its individual bias, with the size of a box representing the magnitude, and the colour its sign. An example Hinton plot can be found in figure 5.9 where the inputs to the network are at the base of the diagram, the hidden layers in the centre and at the top are the outputs. Using this representation it is trivial to discern which weights are contributing to the activations of each unit in this simple network.

Aiding the topological representation of the network was the key goal of Bond Diagrams [WT91], which represent the units within the perceptron explicitly (sized according to their individual bias) and where the bonds between units are depicted varying in length dependent on the weight magnitude figure 5.10a). Although the topological data is clearer, and described within a single plot, the bond diagram suffers

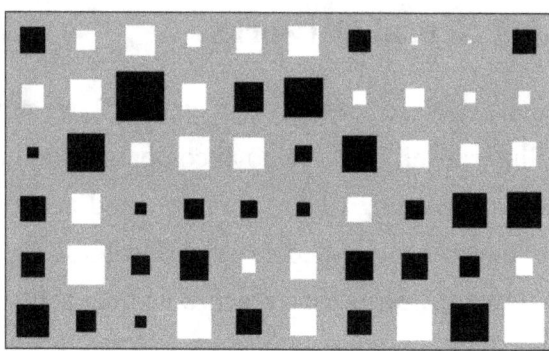

Figure 5.9: Example Hinton diagram, weights are represented by colours for sign, and size of block the magnitude.

from the different forms used to represent bias and weights, and understanding their relative contribution to the activation of a unit.

Further insight can be gained by other diagrammatic forms, including hyperplane diagrams [CS91] where the surface formed describes the linear separability of the problem (fig. 5.10b). According to the number of hidden units within a perceptron layer, the number of divisions within the surface of the hyperplane increments. A similar representation is the Response-Function plots where a diagram for each unit is produced and shading represents the output activation based on the inputs.

The limitations of these static techniques is they cannot represent uniquely all dimensions of the input space, as they are plotted in two dimensions (or pseudo three dimensions by projection), and are of a fixed point in time without displaying temporal information.

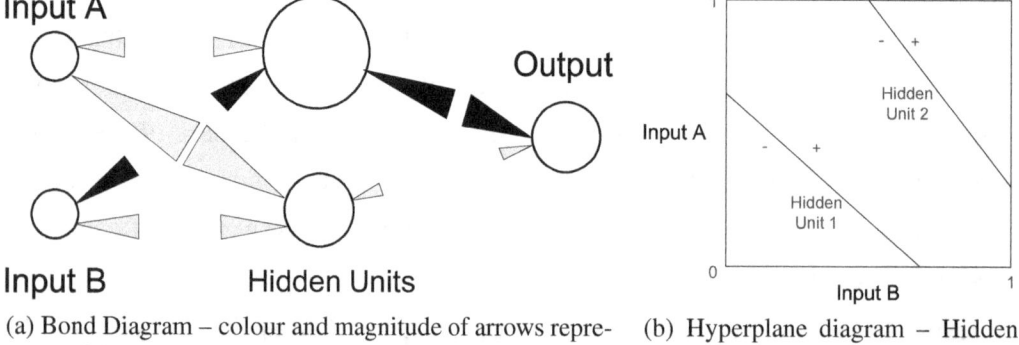

(a) Bond Diagram – colour and magnitude of arrows represent weights, and node-size the biases applied to that node

(b) Hyperplane diagram – Hidden units in the perceptron demarcate the output from the input spaces

Figure 5.10: Bond and Hyperplane static representations.

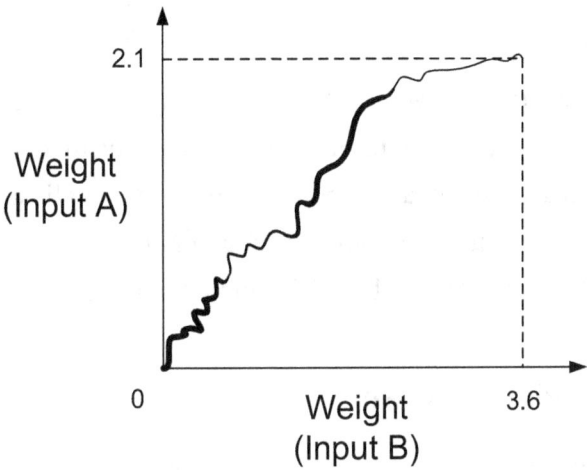

Figure 5.11: Trajectory diagram, weights adjust over time as learning takes place.

**Dynamic Representations**

Static diagrams can either be an instant in time during the training of a network, or are more typically produced at the end of the training period when the network approaches equilibrium and its error minima. However it is possible to animate these diagrams to follow the dynamic adjustment mechanisms of the network during the learning period; examples of animated hyperplanes can be found in [PMKK91, Mun92]. A further plot commonly used which incorporates the time dimension is the trajectory diagram [WT91], which displays a trace of how the weights within the plotted space evolve over the operational period (fig. 5.11). By necessity, the number of dimensions in this diagrammatic form is reduced (creating a non-unique space) to represent the error within the usual gradient descent observed during the learning period. The width and colour of the trajectory may be used to represent attributes of adjustment over time such as the network error or rate of change.

**Reduction in Dimensionality**

As noted, the difficulty in representing multi-dimensional data in a 2 or 3D projection is selecting the data to plot. Principal component analysis (PCA) [Jol02] is used to reduce the dimensionality of the data by decomposing it into a smaller number of components of decreasing variability to the end result. This technique may therefore be used to reduce the dimensionality of a neural network for visualisation on-screen or as input to other software, and is also a useful technique in spike timing analysis, discussed in the next section.

**Further Reading**

This short section has briefly covered some of the major visualisation techniques used in the earlier generations of artificial neural networks (primarily multi-layer perceptrons). Further reading in the area can be found both in visualisation papers [TDM03, UJ06, Rou94, TM05], and in both commercial and non-commercial toolsets [Mat12c, Wil93, KW96, Yos98, SWA01], and on PCA in Gallagher and Downs [GD03].

### 5.2.2 Spiking Neural Networks

In SNNs the output information encoded is represented by the spike timings rather than the continuous outputs of previous generational models. The spike is an 'all-or-nothing' type event; an impulse sent from a neuron along its axon which is imparted to all connected downstream neurons (see figure 2.3). The membrane potential (the current state) of the neuron is decoupled from the downstream neurons in the spiking model, with the information encoded into the existence and timing of the spike events. Within a spiking network there is no clock providing a centralised epoch on which a network is updated, neurons operate in an asynchronous, event-driven manner, their dynamics determining whether they should emit a spike once the membrane potential modelled reaches its trigger threshold.

As the model dynamics are much more biologically realistic within a spiking simulation, there is increased interest for network modellers in visualisation, particularly where the perspective may be user-selected. The visualisation abstractions can be flexible, ranging from single unit dynamics – such as neuron membrane potential over time, through to spike timings of a particular set of neurons, and onto the firing-rates of particular populations of neurons and across the whole simulation. These differing perspectives correlate with single- and multi-unit monitoring techniques plus functional area and whole brain monitoring visualisations that are available in the biological domain, discussed earlier in this chapter. The main advantage of simulation is the flexibility and noise-free granularity offered by a computing system, this permits dynamic scaling from the minutiae of modelled cellular data, to the overview of the activities of the entire ANN, potentially from the same user interface.

**Spiking-Centric Visualisations**

The discrete spiking nature of the third generation of neural networks gives rise to a number of visualisation schemes that centre on the spike events, and their correlation

## 5.2. ARTIFICIAL NEURAL NETWORK MONITORING

with one another. A good proportion of the techniques are based on spike train analysis techniques performed by Awiszus [Awi97], in which he records spike data from one of his own motor neurons, and undertakes a series of analysis on the retrieved data. A number of the more notable spike visualisation techniques derived from this work, and now widely used in artificial neural network study, are now reviewed.

**Raster Plots** Perhaps the best known spike-based visualisation has individual neurons represented in a 'raster plot', a form of scatter diagram where the $x$ axis represents time, and the $y$ axis the identification of the neuron (or series). Each time a neuron spikes it is recorded against its ID and the simulation time, as a vertical bar (c.f. a voltage trace) or 'dot'. Raster plots provide a good overview of all (or a set of) results from an experiment, although it may be difficult even for the human brain to discern any patterns or coincidence with a large volume of scatter data. An example raster plot is found in figure 5.12; this demonstrates the spike trains from a small number of neurons in an ANN over time. The second example has 200 synthetic neurons plotted over a 1 second period (fig. 5.13a). This example illustrates that it is difficult to discern the repeated pattern within a large subset of a population which has the same statistical distribution. In its companion figure (fig. 5.13b) the pattern is highlighted and becomes immediately apparent. The relationship between spike-timings of single, paired and ensembles of neurons can be detected and plotted using several visualisation techniques:

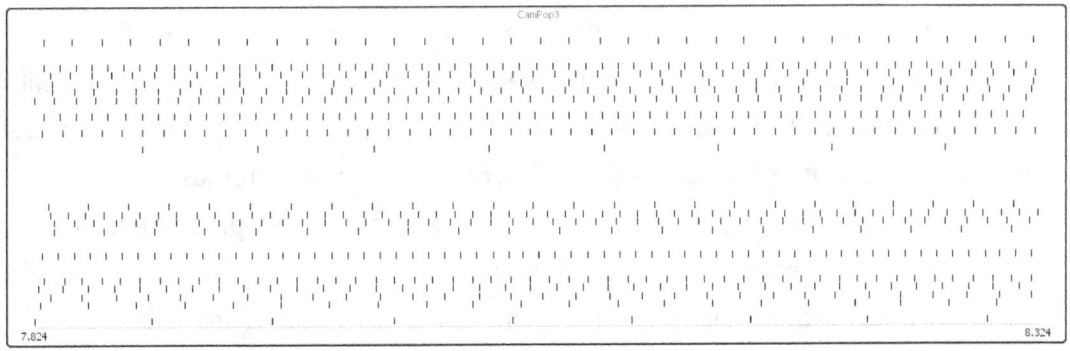

Figure 5.12: Raster plot of spike trains from a number of neurons over time.

**Post Stimulus Time Histogram** This histogram records the time at which a neuron is expected to fire following its stimulation, the distribution being categorised into 'bins' thus providing a probability function of how the neuron will react. By using a pair of neurons and plotting the co-incidence of their spiking in concert, a Joint PSTH

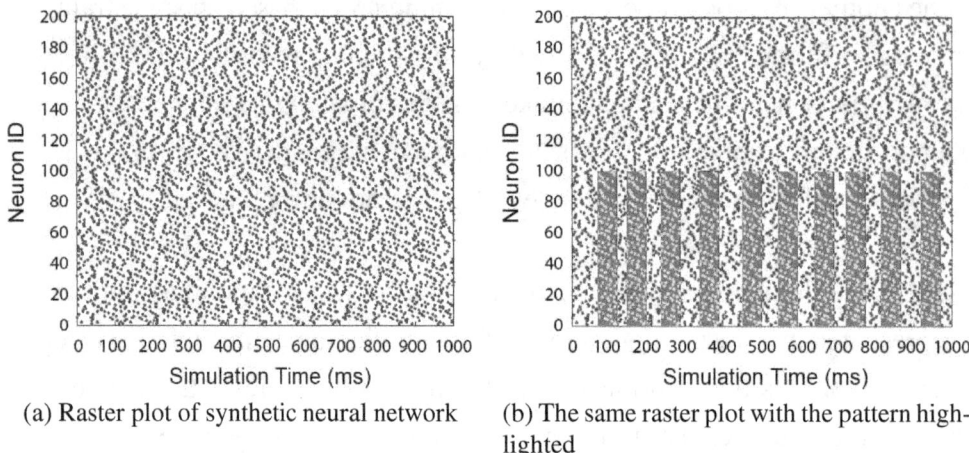

(a) Raster plot of synthetic neural network

(b) The same raster plot with the pattern highlighted

Figure 5.13: Pattern spotting within raster plots. (src: [DRGF11]).

may be formed. Alternatively a normalised JPSTH can be created which removes the independent firing-rates from the correlation and relates just the comparison. Due to their low dimensionality these plots are represented in a grid form, but these methods and presentations do not easily scale to many simultaneous neurons.

**Cross-Correlogram** This, too, is a common technique to examine pairs of neurons, with one neuron's spike train being used as a reference to which a second neuron's spike train is overlaid resulting in a distribution plot over the time period studied. Should the spikes be co-incident within the time bins, or in a manipulation of the spike-train time-base, a relationship can be inferred between the neurons from peaks in the distributions. Clearly the number of pairs of neurons rises steeply in large neural networks and, as with previous methods described, it is most often used with proximate neurons. The auto-correlogram is a variant, where pairs of spike timings from a single neuron are compared (the Inter-Spike Interval (ISI)), and this analysis may be used to pick out rhythms and other characteristics of the firing of that neuron or sets of neurons. A survey paper detailing the challenges of the analysis of spike-train data (particularly of a multi-dimensional nature) can be found in [BKM04].

Not all data of interest in ANNs is based around the spikes. In computational neuroscience [SKC88] neuron and synaptic simulations are represented by mathematical models. Being within a computer model ensures that the state of all constituent elements may be queried, at any time, and in a noise-free repeatable environment. This data may be useful to the user, and may be visualised too.

## 5.2. ARTIFICIAL NEURAL NETWORK MONITORING

**Non-Spiking Data from SNN Networks**

As with earlier network generations, weights remain important in spiking neural networks, and are used where input spikes arrive at modelled synapses. They determine the post-synaptic influence of incoming spikes on the downstream efferent neuron. Weights may be altered in simulation by plasticity models, e.g. Long-Term Potentiation and Depression (LTP and LTD), and Spike Timing Dependent Plasticity (STDP). Learning rules such as these are instances of Hebbian Learning (chapter 2), which is commonly paraphrased to 'Cells that fire together, wire together'. Monitoring of overall strength of individual and groups of synaptic weights over time is a common case for study in spiking ANNs, just as in previous generational models.

**SNN Visualisation Software**

Many solutions created to visualise spiking neural networks are bespoke, tied to the implementation software used to specify and execute the simulation. Several of the tools also permit interaction with the model, as it runs, to change parameters or adjust settings for the next simulation via GUI software to aid the modeller. This section summarises the visualisation features in some of the more common neural network modelling software packages (section 2.3.1), and further software available in this area.

**Visualisation in Popular Neural Modelling Software**  Some of the more popular software modelling tools were listed in 2.3.1, and examples of several of their visualisation output capabilities are in the 2007 paper from Brette et al. [BRC+07]. This paper provides examples of many of the visualisation modalities across several tools, including NEURON [HC97], GENESIS [BB07] via XODUS, the Neocortical Simulator [Bra12a], NEST [GDG07] and others such as (P)CSIM ((Parallel) Circuit Simulator) [PN09] which was previously seen used as a simulation engine for PyNN [DBE+09] in section 2.3.1. The Python oriented simulators such as PyNN are able to make use of the libraries available in Python to perform graphing, and one notable example in this space is the NeuroTools suite which provides a plotting environment oriented to the needs of neural network simulation [DBE+09]. Brian [Goo08] is also Python based, and has two custom plotting environments for raster and histograms plots, with the remainder of the plotting expected to use the functions from MatPlotLib [HDD11] which is also the basis for the NeuroTools plotting routines.

**iRaster** iRaster [SSSB10] is an analysis tool for post-simulation spike-train data, and renders raster plot data in sliding windows as users are familiar and comfortable in exploring and identifying patterns within this representation. It is able to recover and display a subset of synchronously active neurons detected using a number of techniques including reordering and visual manipulation of relevant spike train data. The GUI provides facilities for multiple views of data, and facilitates users zooming in on 'items' of interest. The tool may also use synthetic or biologically sourced data, and is able to inter-operate with various standard sources of spike data, including in MATLAB binary form – the authors reasoning that another incompatible format will limit the iRaster's exploitation. The main advantages of the software are this wide-ranging support of data sources, although it does appear focused (limited) to spike trains and their correlations only, and is not able to represent other forms of input data.

**MATLAB** Many tools use MATLAB [Mat12a] to provide their input and analysis / visualisation functions – indicated in section 5.1.4 and a key finding of software and resources sections of the NeuroDebian survey [HH11]. MATLAB may be used for both biological and simulation data, and thus provides an avenue for tool-sharing, although this is rarely the case practically. One example of a Spiking Neural Network toolbox for MATLAB is detailed in [BMT09]; this allows for a full-cycle of description, operation and visualisation of SNN networks. This implementation builds a layered network around models of neurons (MATLAB being highly suitable to explicit description of the mathematical dynamics of neural and synaptic models). The visualisation technique uses a multi-window environment specified by the user to view the dynamics of the network, from the neuron and its parameters through to spike-rates which are better suited to large network sizes, and to rate correlations. MATLAB Central [Mat12b] is an ideal place to begin when examining the capabilities of MATLAB to perform and analyse spiking neural networks [Izh03]. MATLAB has the advantages of being easily extensible and having the manipulation tools and libraries to perform analysis and signal processing of results data, and that many users are already familiar with the tool for other analytic purposes. It has the disadvantages of lacking a standardised spiking neural toolbox, and its lack of integration with the emerging standard of PyNN based descriptive specification.

**Nengo** Nengo integrates a highly graphical environment for analysis and evaluation of the networks built using the NEF (Neural Engineering Framework). Due to this

## 5.2. ARTIFICIAL NEURAL NETWORK MONITORING

tight intertwining there is a strong emphasis on visualisation and parameter interaction with in-flight networks. For smaller networks in Nengo it is possible to operate and interact with the networks created in real-time. However, in larger networks where this is not possible, the simulation runs at a slower rate, but it still remains possible to interact with the simulation parameters. Plots can be dynamically created directly from the GUI interface (see figure 5.14 for examples, including voltage levels, raster plots, tuning curves, and grids of voltage levels and activity rates). The environment is created in Java providing some level of platform independence, extending to incorporation within the MATLAB environment if the user wishes. External renderers may be used to incorporate visualisations that are not built in, and data may be plotted with a '3D' flavour using this technique. The GUI is a flexible best-of-breed visualisation experience, however the NEF is not a generic solution for all neural modelling (and therefore visualisation).

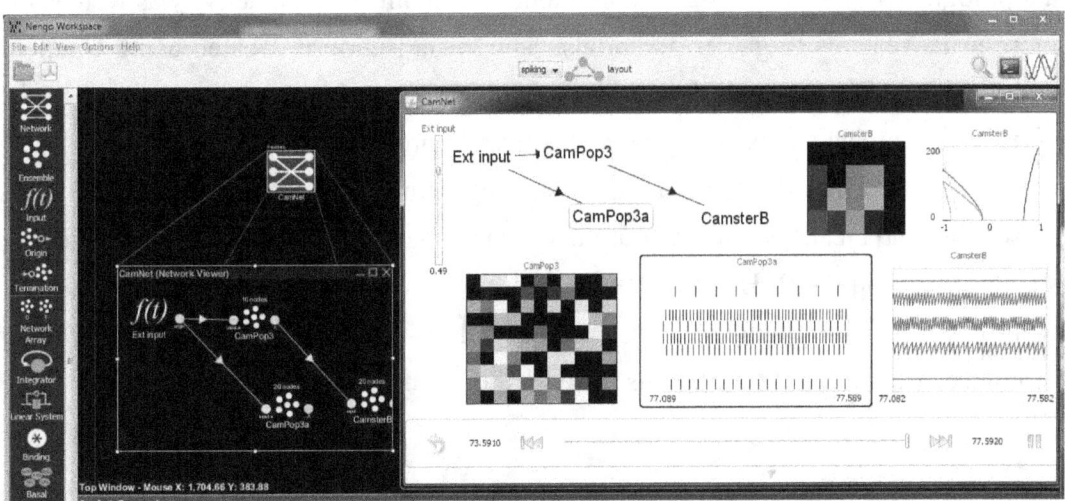

Figure 5.14: Screen taken from Nengo with multiple population based plots displayed as the simulation operates. In-flight interaction via input function slider is also illustrated (in the figure the input voltage applied is 0.49V).

**Neurogrid**  Neurogrid also includes the ability to visualise within a real-time simulation environment [GLS+05], their web pages indicating that it is possible, from a GUI, to select different granularity within the network to visualise – from the low-level individual neuron voltage through to raster plots from a modelled cortical layer. Neurogrid provides a layered view, this perspective is derived from the topology of Neurogrid networks which are constructed in hierarchical layers. Although appearing less flexible than the Nengo interface, different plots may be combined within a

GUI window. These cover raster plots, rate plots and the voltage levels of monitored variables – the ability to present various modalities of data being a distinct benefit to the user. Audio, too, may be used to aid comprehension of network operation, with clicks sounding as a neuron or group of neurons fire, the frequency of which indicates their activation. The main disadvantage of the this tool is that it is tied closely to the Neurogrid spiking hardware and is non-generalisable.

**SpikeFun**   One amateur project SpikeFun [Dim12] uses a single desktop processor to model over 3 million neurons and 50 million synapses. It is not able to operate in real-time for larger simulations, but may perform simulations over regular and 'brain shaped' systems. Its visualisation capabilities are of interest, however, as they provide a topological view of the system being modelled, and its current activity as well as the more usual features to gather potentials. The visualisation is created using OpenGL and thus allows easy manipulation of user viewpoint, and slices to be taken from the system to examine status at any point (as the 3D projection naturally obscures some detail). In its favour the tool has some striking demonstration material available via its website and creates simulated output similar to biological observational techniques such as EEG and fMRI. However its main disadvantage appears to be the lack of practical experimentation facilities – it would be interesting to tie this tool in with other biologically significant simulations to study their behaviours graphically.

**SNN3DViewer**   [KPP09] This tool concentrates on the 3D representation of maps of neural networks, permitting users to explore the dynamics of a system (post-simulation) with a network-centric perspective. In this way the activity of a network may be visualised with individual spikes travelling over axons and dendrites in the network, with colour representing the strengths of connections. The sheer number of connections in large networks may make visualisation difficult, and it is possible to select the connections of interest for display and to disregard the others. As with all 3D projections data may be obscured and SNN3DView contains controls to adjust the viewpoint to get a better overall picture over time. Another disadvantage of this technique is that real-time performance is not possible as it would be too fast to display, therefore this tool has largely aesthetic appeal.

## 5.2. ARTIFICIAL NEURAL NETWORK MONITORING 117

**Co-registration / Fusion**

Within the SpikeFun tool a prototype of a brain is used, and plausible activity projected onto it. This technique is similar to the activity sequences from biological sources such as functional MRI data which are plotted in conjunction with structural scans (co-registration). If populations of simulated neurons may be mapped to functional areas then a pseudo-fMRI fusion visualisation is created. This technique has been used within the Blue Brain project, albeit not across an entire brain (see the project presentations [INC08, Con11]), and has been used with large Izhikevich ANN models, where activation is projected onto a template human cortex [IE08]. As larger biologically plausible networks come into existence, exploiting the co-registration technique is a powerful method of data visualisation and results analysis. If the data can be collected and stored in the same format as the biological data, then the analysis tools can be shared and bridges built between computational and wet neuroscience.

**Themes from Spiking Visualisation Tools**

The visualisation schemes used across the Spiking ANN tools include many of the same types of representation seen in the biological, and non-spiking ANN imaging spaces. The data types commonly comprise: raster (scatter) plots, histograms and current / potential plots over time. These diagrams are often multichannel, as the points of recording are numerous, and they support visualising data using more than one method or viewpoint. 3D and tomographic representations of neural data, although fascinatingly organic, serve the purpose of discerning high-level patterns of behaviours such as oscillations. However it is the ability to 'zoom' into the detail of the implementation, all the way from the '40,000 feet' overview to populations of neurons and their firing-rates and patterns, and then again to the individual neuron and to its dynamics that truly provides the user with a valuable experience.

There are few, if any, standardised tools which operate in the simulation space; each simulation tool either bundles analysis tools, or the output is recovered and analysed in tools such as MATLAB off-line, using custom visualisation techniques developed or adapted for the specific application.

The majority of neural network results analysis (biological and artificial) takes place off-line, and there are several reasons for this. Firstly the quantity of data recovered can be vast – and creating a channel for this data back to the visualising software is challenging, particularly as most of the data is discarded as it is irrelevant from the users viewpoint. This channel itself may be limited in capacity, or temporal accuracy, providing the equivalent of peering out of a frosty window although this may be sufficient to provide the results required. By using an off-line dump of all data collected from the network, analysis tools may drill directly to the required information in the system but this analysis is delayed until after the simulation is complete, and by the time taken to obtain and process the data.

## 5.3 Summary

This chapter has explored techniques used in brain scanning for its functionality and structure, techniques which can be applied in research and for medical purposes. Monitoring can be active (by invoking a physiological response), or passive (by external sensors) which is less likely to be injurious to the subject. There are several types of active monitoring techniques, with a clear distinction made between those which involve radioactive materials, (and are therefore limited in their repeatability due to the risks of cumulative radiation dosing), and the alternatives. Techniques such as MRI (based on magnetism) and NIRS (based on light) invoke a physiological response to the applied stimulus and are thus regarded as active, although no adverse side-effects of these techniques have been determined to date. In all brain imaging techniques the resolution of the information retrieved is limited by the sensor technology of the scanner and its situation. It is only by using invasive or in-vitro techniques that specific cells rather than imprecise functional areas within the brain can be studied.

Computational neuroscience has therefore developed models of the biological constituents of the brain which may be studied in-silico, repeatedly, and non-disruptively in a noise-free environment and at any level of abstraction. As the models developed may emulate individual neuron function, it is realistic to produce and image single-neuron data totally flexibly, which is an increasingly active area of research in both wet and computational neuroscience.

As SpiNNaker targets primarily third generation simulations: Spiking Artificial Neural Networks, the biological and artificial visualisation methods described by this chapter may be appropriately applied to the simulations SpiNNaker runs. Chapter 6 goes on to describe the flexible visualisation framework developed to support the management of real-time neural network simulations (from SpiNNaker and beyond).

# Chapter 6

# Visualising Neural Networks on SpiNNaker

SpiNNaker has been designed with the primary purpose of simulating real-time spiking neural networks [FB09]. This is an unusual approach to take within the artificial neural network space as, in addition to the requirements of fidelity and scale, a rigid timing constraint is added to the system requirements. If a real-time system does not meet its timing then its results may be invalidated if the constraints are hard, or at best compromised within a soft constraint environment.

Artificial neural networks operating in real-time, however, open the opportunity for interactions with real-world objects and systems. This gives rise to applications where the feedback loop may be closed, with the network interacting with sensors and actuators to provide a real-time control system. In such networks time-constraints may take on more significance than in pure simulation, as the tangible system being controlled may be fragile or have responsibilities with consequential importance – and may be expected to operate effectively over extended periods of time.

Three factors bound the operation of artificial neural networks on a system: the size and topology of the modelled network, the computational complexity of the modelled elements, and the time-constraint on its operation. A real-time approach for simulation is interesting, as all neural networks in biology operate in real-time, so the size and fidelity of spiking simulations must be selected to meet the constraints of the target platform. SpiNNaker, as a target platform, is a high-performance architecture designed to meet the requirements of real-time operation of neurons which spike in a biologically consistent manner. In its largest configuration SpiNNaker may model 1 billion neurons with a (gross) mean firing-rate of 10 Hz distributed to 1,000 efferent neurons.

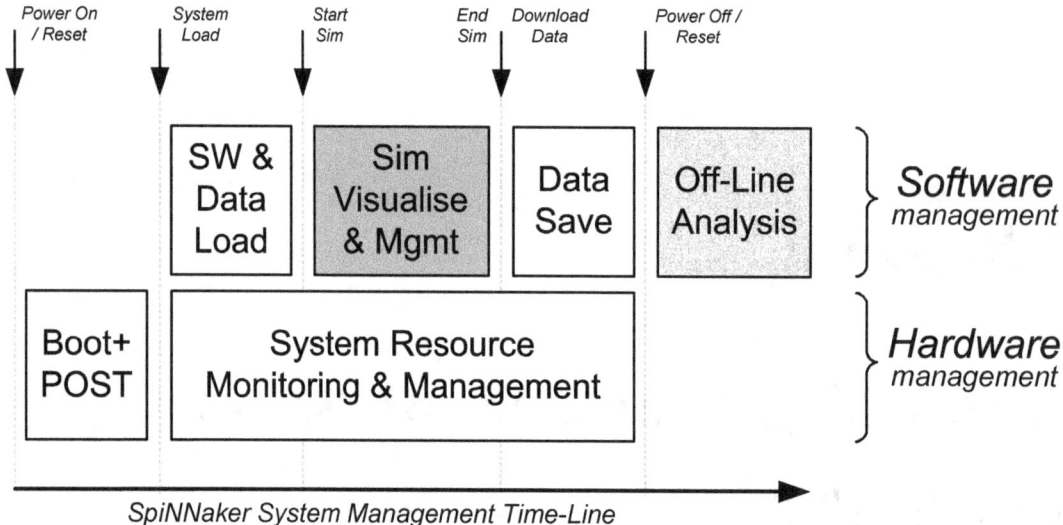

Figure 6.1: The system management time-line of the SpiNNaker platform – software management is performed by the visualisation software. (The visualiser is also capable of performing limited off-line analysis – hence the hatching here).

This chapter explores the real-time visualisation framework that has been created to support the management of SpiNNaker's ANN modelling software. Large scale networks generate vast quantities of data, and the challenge is to monitor the relevant data set from the system, without overwhelming its network or processing capabilities. A visualisation scheme for SpiNNaker needs to be able to meet the requirements of the user – to present data in a clear and consistent form – but also be scalable. In SpiNNaker's case the range of visualisation scaling starts at individual neuron dynamics passes through populations of neurons, and onto an overview of the network itself; all while it is operating and without degrading the simulation itself.

A successful management visualisation tool permits users to examine behavioural patterns in real-time data from the explorable system areas. Within the management time-line this function operates as highlighted in figure 6.1 and additionally allows interaction to influence the behaviour of the operational software.

## 6.1 Principles of Visualisation

Whilst many of the fundamentals of scientific visualisations were described and referenced in the previous chapter, it is the transformation of neural data into a graphical form to aid users' comprehension that is fundamental to the successful implementation of results visualisation. There are many techniques for representing data, from glyphs to histograms and surfaces to video rasterisations, but what makes a good visualisation in the ANN space? Replicating the tools from the previous chapter would benefit from the experience already within the field, and users will be familiar with them. With visualisations being created in real-time and dependent on computer graphics, user interaction with this environment is key to the best use of the visualisation experience. An oft cited seminal paper from Schneiderman [Shn96] explores the area of user interaction with rapidly expanding sets of data to present a coherent and easy-to-use visualisation interface. Underlying this paper is the visual information seeking mantra: 'Overview First, Zoom and Filter, then details on demand'. This simple set of visual design guidelines is surprisingly effective; to begin with an overall picture and then to zoom into the details that are pertinent. This forms the backbone of the visualisation of and data presentation for SpiNNaker software management.

## 6.2 It's All Just Data

In the previous chapter visualisation techniques used in both biological and artificial network monitoring were examined. The goal of the visualisation software is to provide this type of functionality for the SpiNNaker platform so that users may manage their in-flight simulations as they operate on SpiNNaker. In this way simulations do not appear as a 'black-box' to the user, or merely run as batch jobs with analysis taking place off-line following the completion of the job. However it is noted that the majority of visualisation modalities used: scatter diagrams, histograms, voltage traces and grid plots are not specific to particular neural parameters – they may be used many times over, to represent many types of data generically. Therefore the approach taken was to plot the data independently of what it actually represents – and for each simulation a simple module is prepared which interprets incoming data, processing it into a form that may be plotted by whichever core visualisation modality is selected as most appropriate by the user.

## 6.3 Visualisation Targets

The target of the real-time visualisation platform is to be able to explore operational neural networks across a range of abstractions from cell dynamics through to population activity, as simulation data changes in real-time. This live monitoring, particularly in long simulations, may be advantageous if it presents enough information to the user (or computational agent) to make decisions and adjustments interactively, or if the situation is not recoverable, to terminate an unsuccessful simulation early to save resources.

Large ANN simulations are capable of producing prodigious quantities of data, and it is clearly not practical to replicate this to the visualisation engine(s), nor would it be feasible to represent all this data in an comprehensible form. Therefore visualisation software should take a layered approach to the provision of its real-time data, as when visualising their networks, users will typically wish to monitor a particular segment of its behaviour. To enable this viewpoint data is filtered and aggregated both internally within the software framework of the system, and externally – dynamically turning the monitoring capabilities on and off as required by the user.

If too many sets of information are requested, the data network or the resources of aggregation points or the visualiser tool itself may be become swamped. A trade-off needs to be made to observe greater detail at a small scale, or to have more data points but with a lower resolution. The bandwidth available for real-time monitoring is never going to match the bandwidth of data available to tools in post-simulation analysis – therefore this live aggregation remains key to the implementation.

While implementing the real-time visualisation tool an attempt has been made to provide a general extensible interface which can be used over a variety of neural network models, and one that places little or no burden on the hardware performing the neural simulation. A variety of visualisation options have been provided, and real-time simulations on the SpiNNaker system have been used as the test platform.

## Communicating Data for SpiNNaker Visualisation

SpiNNaker, as a computer architecture, is a flexible memory-mapped system that may be probed at all levels, from hardware through to software. Although its communications networks have been sized to cope with the demands of a biologically plausible set of interconnected neurons, non-spike management and control-type traffic for the visualiser also uses this shared infrastructure.

In SpiNNaker, non-spike traffic differs from (the majority) spike traffic in that instead of flowing from one-to-many it typically flows from many-to-one. Details of the available packet formats used by SpiNNaker were covered in section 3.4.3 and illustrated in figure 3.4. At first glance, it appears that Fixed Route (FR) packets are most suitable for management purposes, but using FR packets requires that each Application Processor be programmed to create them. This is not an ideal situation, as the

## 6.3. VISUALISATION TARGETS

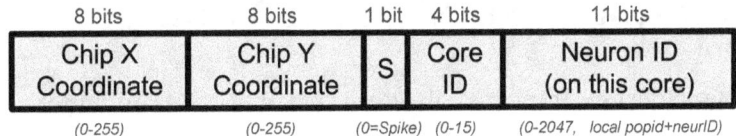

Figure 6.2: SpiNNaker spike packet key format (used throughout the machine).

Application Processors should be performing neural processing, and one of the stated goals for the real-time monitoring is that it should not impact on the neural processing. Point-to-Point packets similarly require specific support and are routed between Monitor Processors only; Nearest Neighbour packets are not suitable as they are passed between Monitor Processors and would require hop-by-hop software routing.

For visualisation, therefore, advantage is taken of multicast routing and specific aggregation cores throughout the machine (typically co-located with Ethernet connections) which act as collectors for information in the system. If the collectors are capturing information regarding the spikes then simply by adding the collector node to the distribution tree of the monitored neurons it acts as a passive tap. For other types of data collection it is useful to visit the multicast packet key field format (fig. 6.2). The 32-bits in the multicast key-space are arbitrary, though by convention the first 16-bits are used for origin (8-bit $x$ and $y$ Cartesian coordinates), followed by the 'S' bit representing Application (0) or System (1) space. Within the application space (shown), the remaining 15 bits represent the application core and the firing neuron ID. If the 'S' bit is 1, there are 7 system categorisation bits (representing management / synchronisation / experimental etc.), and the remaining 8-bits are free for user-defined signalling (plus an optional 32-bits of payload if required).

By loading routing tables appropriately, the system space within the 'multicast' key space may be used for destination-based routing and monitoring purposes. The targets of this management traffic are dedicated collector or aggregation cores such as the 18th core on fully-functional chips, and multiple overlapping management trees may exist concurrently (unlike FR). By using in-network collation of visualisation data, aggregation and filtering may be performed *within* the network, rather than by transmitting the data externally, thus saving external processing and networking resources. As an example, in a large neural network where there are many populations of neurons, an aggregator node can capture data from the neurons in each population, and calculate a spike-rate for that population, and periodically send this aggregated information to the visualisation platform for display.

## 6.4 Implementing SpiNNaker's Visualiser

A schematic of the design of the visualisation system is given in figure 6.3. The visualisation software operates as a number of threads to maintain modularity and so that the separate tasks may operate independently. Aggregated live data is fed from the neural network simulator through the network to the packet decode section of the visualisation software. This strips the implementation-specific headers from the packet so that the data payload may pass into the data-specific visualisation module and then to the visualisation engine. To the right of the figure an optional recording feature is provided which allows later replay of data for testing and further analysis if required. The following sections provide brief implementation details for each of these components which together form the visualisation management system.

### 6.4.1 Execution Environment

The visualisation software is built using C++ to perform data manipulations and meet the network and file I / O performance requirements. The visualiser makes use of threads to partition the execution of the modules (as indicated in figure 6.3), ensuring that each thread may not block the operation of another. The test systems are GNU / Linux platforms as this is the prevailing operating environment used in the computational neuroscience community [HH11]. The visualiser has been tested and executes directly on contemporary Ubuntu and Fedora operating systems and, using a Ubuntu VirtualBox, on an Apple MacOS X host machine. For the graphical presentation OpenGL was selected as a platform-independent toolset to take advantage of the available graphics hardware to improve performance. The GLUT / FreeGLUT libraries are used in the compilation process for this. OpenGL also simplifies the construction of the visualisation environment and window management so that the code remains platform independent.

### 6.4.2 Packet Decode

A modular approach is taken to the interpretation of data arriving via the network connection. Sockets are opened, as necessary, to receive input data and protocol specific structures used to gain access to the data therein. This structure may be reused across different simulations from the same platform, with only separate visualisation modules for each different visualisation requirement. For example, the decode structure

## 6.4. IMPLEMENTING SPINNAKER'S VISUALISER

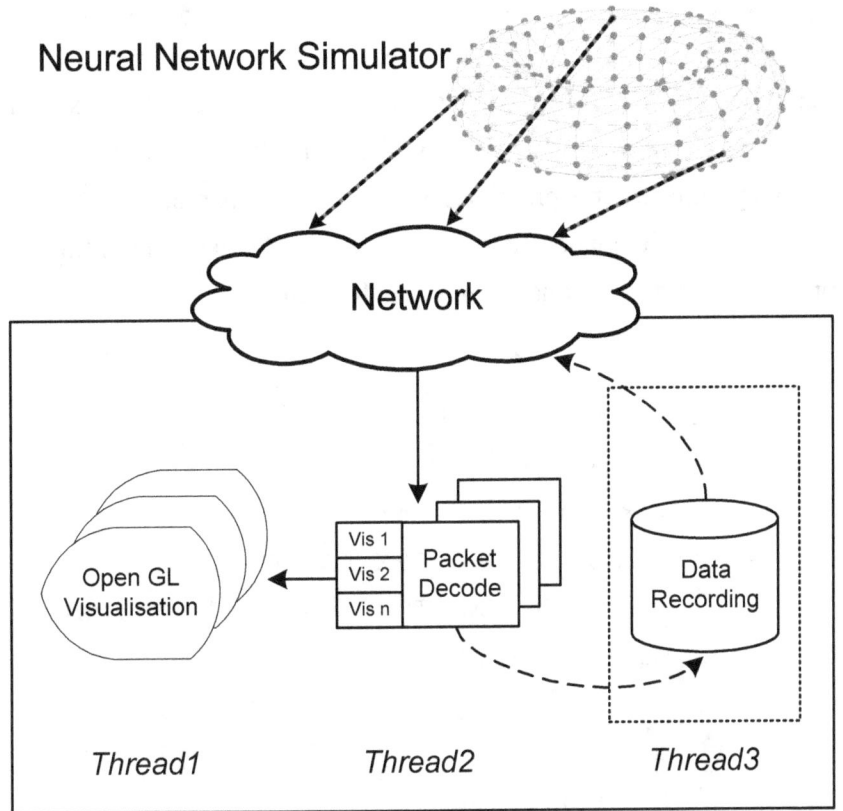

Figure 6.3: Schematic of the visualiser operation, data arrives over the network from the simulation platform and is decoded, then filtered / transformed into the format required by the specific visualisation. Data received may optionally be recorded and replayed into the visualisation for post-simulation analysis.

supporting the SpiNNaker SDP packet format (fig. 3.7) is 16 lines long, and for the basic SpiNNaker packet format (fig. 4.7) – as also used by the flood fill routine – is even simpler at 7 lines.

**Visualisation Data Storage**

The data structures used internally within the visualiser are in 2 forms as data may either be represented based on the last set of data received: *'immediate'*, or it may temporal in nature and have: *'history'*. The data is stored in dynamic array-type structures to allow adequate storage to be reserved to meet to the requirements of all data that may require visualisation. Data may also be represented in a variety of formats, from integer to fixed-point to float, and the user may specify this per simulation.

**Visualisation Decode**

Once a packet has arrived and its packet decoded it will pass into a specific visualisation decode stage as each visualisation will usually have a different format of data presented, or may require some pre-processing before being placed into the appropriate positions in the 'immediate' and 'history' data arrays. An example of this visualisation module for a simple 2 dimensional example is found below.

```
1   xsrc=sdppkt->srce_addr/XMAX;
2   ysrc=sdppkt->srce_addr%XMAX;
3   for (int i=0;i<numDataItems;i++) {
4      uint arrayindex=(EACHCHIPX*EACHCHIPY)*((xsrc*
                    (XDIMENSIONS/EACHCHIPX))+ysrc) + i;
5      if (freezedisplay==OFF) {
6         immediate_data[arrayindex]=sdppkt->data[i];
7         history_data[time][arrayindex]=sdppkt->data[i];
8         somethingtoplot=TRUE;
9      }
10  }
```

Lines 1 and 2 take the sending address and derive the $x$ and $y$ co-ordinates of that node (a packet may only have one source, but the internal data may present further categorisation information – such as core or neuron IDs). Line 3 sets up a loop to read all the additional words of data from the data packet. Line 4 calculates the position in the storage array that will be populated (here based on the physical dimensions of the system). Line 5 detects whether the user has paused the display and hence the data should not overwrite the immediate paused data. Line 6 loads the immediate data from the received packet and line 7 loads the appropriate position in the circular history buffer with the same. Line 8 updates a global variable to inform the visualisation module to update the display at the next opportunity as new data has been received. Not shown: The historic data is overwritten with null data as it becomes stale as part of the visualisation code. Pre-processing may also be applied here, for example calculating rolling-averages, incorporating decay, performing integration or other numeric manipulations as required. As can be seen, adding a new visualisation for a specific neural network requires relatively little work on the visualiser.

### 6.4.3 Visualisation Interface

This section provides a brief overview of the implementation of the visualisation interface presented to the user and its features. The following facilities are available for the visualisation of the immediate and historic data arrays. Examples of the output from the various visualisation modalities can be found in the results sections, so are not explicitly provided in this section.

**Modalities**

In the previous chapter an introduction to some of the methods used in both quantised biological data and in artificial neural networks was given. These, and requests from users, led to the following visualisation techniques being implemented for use for real-time SpiNNaker software management:

**Temporal Displays** This visualisation modality includes the ability to plot values and data over time. The user has the ability to alter the window duration for the historical data using keyboard input, and may specify a default at simulation definition. The start of the temporal displays is triggered by the receipt of the first piece of data, so the user may begin the visualisation software in advance of loading the simulation software onto the target system. The visualiser supports the following temporal plots:

- The raster plot. This plot represents each incoming event received for a channel (monitored item), and either uses a vertical bar or dots depending on the number of items being plotted and the size of the plotting window. For non-spike data this is useful to view the timings of input data items for a channel.
- Line diagram. This diagram plots output channel values over time as lines for each channel, with the $y$-axis representing the quantity being plotted, e.g. the membrane potential voltage. This variant plots all items on a full-height plot meaning that series may collide or intersect.
- Multichannel diagram. This second variant of the line plot is similar in output to an EEG, with the $y$-axis split into discreet sections for each channel being plotted. The colour of either of the line plots may be chosen to be constant based on the channel number, or dynamically adjusted based on the last received value.

**2D Plots** A number of constantly changing 'immediate' displays based on last received values may also be chosen by the user if preferred, with data displayed with the

$x$ and $y$ coordinates required. This is often useful for topologies based on the physical layout of the simulation.

- Tiled. This is a block 2D plot, with the colour of each rectangular tile in a grid changing based on the last received value from that channel.
- Interpolated. A variant of the tiled plot which creates a pseudo-node at the intersection between adjacent tile corners, and interpolates the value linearly. Smooth graduated triangles are used around the pseudo-node to the actual channel data points, and can make for some very smooth and attractive plots, especially where there are smaller numbers of related channels. The data however is interpolated and not a true-representation of the channel data.
- Histogram. This is similar to the tiled plot, but where the number of tiles in the $y$ dimension is equal to 1, and for each channel there is a bar along the $x$-axis. The height of the bars in the histogram changes dynamically with respect to its channel value.

It is possible in most plots to spawn multiple windows with the view of each based on a sub-set of the zoomed and filtered data, or to open a new window with an alternative view of the same data set.

**Features**

As well as the choice of visualisation modality, the users have a number of other features to choose from in the visualiser. These include choosing the colour-map to be used for representation of a range of data values, resizing windows, going 'full-screen', printing data values as text, and transforming the data with rotations and flips as the user desires. Options may be chosen by keyboard short-cuts, or by a right-mouse-button click menu. An example of the visualiser and its menu system can be found in figure 6.4.

One other notable feature allows the user to save data from the visualiser as it is received (fig. 6.3) in a number of proprietary and standard formats. The user may then choose either to increase or decrease its later playback speed as required. The reason this option is hatched in figure 6.1 is that the temporal accuracy of this saved data is dependent on the network / disk performance as it arrives at the visualisation management station and on any aggregation performed. Users are therefore encouraged to perform detailed post-simulation analysis on data saved internally to the simulation if available.

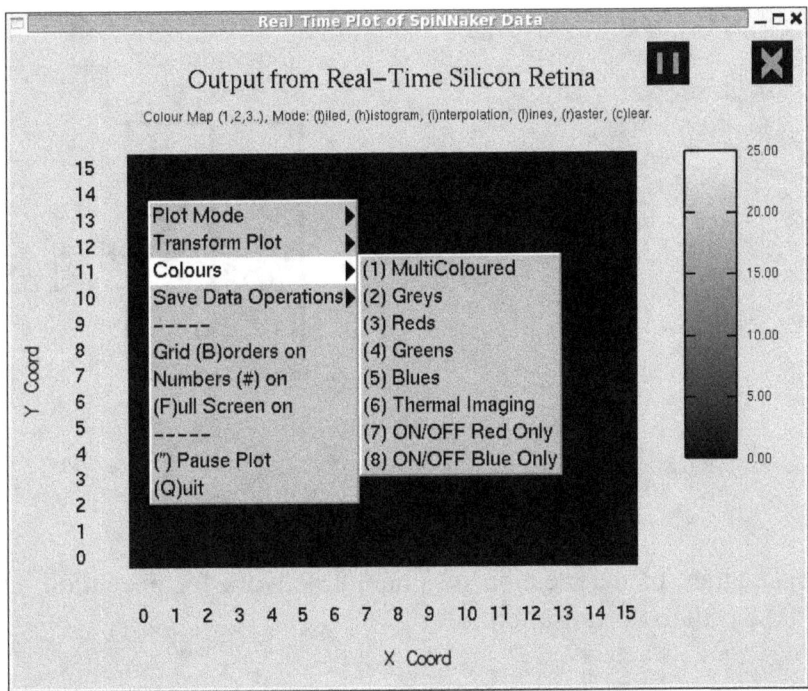

Figure 6.4: Using the mouse right-click brings up the visualiser's menu system – here the colour map / key menu is being expanded.

### 6.4.4 Mapping

When creating neural networks for simulation on SpiNNaker, the high level description of the network is mapped down onto the SpiNNaker hardware using the PACMAN process (section 3.5). Populations of neurons may be split and dispersed between multiple cores and chips to achieve efficient use of resources. The SpiNNaker chips are unaware of this mapping process, and the routing takes care of ensuring the continuity of the network design.

During the mapping stage, each neuron is assigned a global neuron ID within the system, formatted as in figure 6.2. This neuron ID is formed from the physical coordinates of the chip, the core ID in that chip and a local neuron ID. The local neuron ID is formed from a variable length *local population* which, if a core has neurons from more than one population, differentiates between them and concludes with the neuron ID within that local population.

In the mapping system illustrated in figure 6.5, the high-level network model description passes through the partitioning and configuration manager [GDR+12]. This system takes into account the type and dimensions of the target system (e.g. SpiNNaker), and maps the network into a configuration to be executed.

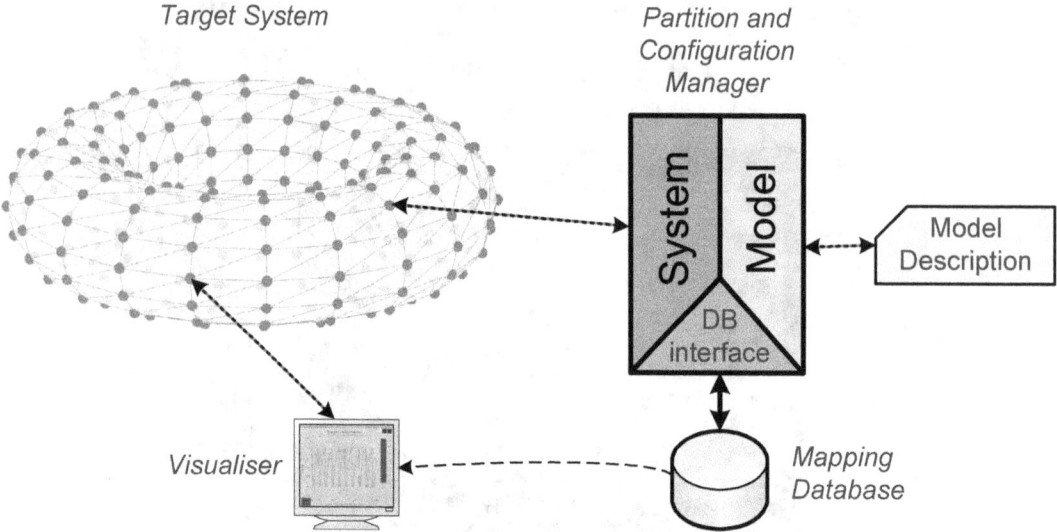

Figure 6.5: Diagram of the creation of a neural network for execution on the target system, and the path to visualisation.

Therefore, information created in the model-to-hardware mapping process must remain accessible to the visualisation software to re-create population-based views from data received directly from the hardware. The visualiser is able to access this local-to-global mapping information and use this in the organisation of data presented on screen and to consolidate diversely located populations.

### 6.4.5 SpiNNaker Neural Software and Accessing its Data

SpiNNaker is an event-driven, scalable, programmable platform and targets the real-time ANN operating space. It may operate anywhere between the high fidelity simulations used in projects such as Blue Brain [Mar06], whose approach is to use biologically plausible Hodgkin – Huxley type dynamics [HH52], through to simplified models which target only the most basic behaviours such as Leaky Integrate and Fire (LIF) [Ste67] and other 'computationally light' models such as Izhikevich [Izh04].

Test boards (fig. 4.13) with four SpiNNaker chips (72 cores) have been available since May 2011 and these 4-chip SpiNNaker simulations can support many thousands of simulated neurons and synapses, with modular 48-chip boards delivered in June 2012 increasing this twelvefold. Results may be analysed post-simulation by downloading data stored within the machine to tools such as NeuroTools [DBE+09], MATLAB [Mat12a] or gnuplot [WKM11]. It can take a significant time to transfer the contents of a simulation to a Host machine for analysis, for example 4 SpiNNaker

## 6.5. RESULTS

chips contain 512 MB of RAM. The Ethernet connection provides for a few MB / s after overheads and therefore it can take over a minute on a 4-chip system to transfer the required data for analysis.

Whilst it is therefore impossible to send a full-set of data on the real-time output channel, within a large spiking neural network the user may choose the level and precision of the data to make best use of the resources, e.g.:

- Observe few neurons with a good precision: Neuron state variables
- Observe a population with spike precision: Raster Plot
- Observe a larger area with more abstract information: Mean activity rates

The same channel capacity applies regardless of what is being visualised; here it is a trade-off between close visibility and great detail, and wide visibility with less. It is the switchable aggregation and collation of data within the system that enables this to occur.

## 6.5 Results

Results are presented from the implementation of the SpiNNaker real-time visualiser, whose implementation follows the approach described by this chapter. As SpiNNaker is a programmable architecture supporting different neural models and network topologies, the visualiser must be able to represent data with different levels of abstraction, and organise it in conjunction with the logical to physical mapping information found in the PACMAN database. The visualisation tool offers diverse visualisation representations, so the users may select the most apt for their chosen incoming data. The examples presented demonstrate the visualisation software's flexibility, rather than intricacies of the neural models which are not the focus of the management work in this thesis. The results cover a number of neural network visualisation examples, across a range of abstraction levels: from single neural dynamics to aggregated population measures.

### 6.5.1 Neural Dynamics

Neural networks are composed of single elements – neurons – which are often modelled as being governed by sets of differential equations. Such equations describe the dynamics of the internal states of the neuron, for example its membrane potential. It

# 132 CHAPTER 6. VISUALISING NEURAL NETWORKS ON SPINNAKER

is of interest to observe such internal states to verify the correct implementation of the model's atomic elements and to compare them with single-cell recordings made from the biology. It is also possible to use the same approach to observe synapses, for example the current injected in at the neuron, or the change of a synaptic weight over time in a plasticity model.

Figures 6.6 and 6.7 show a practical use of the visualiser at this level where it can be used to detect implementation errors in neural models. A current based LIF model was implemented whose dynamics follow equation 6.1:

$$\tau_m \frac{dV}{dt} = E_L - V + R_m I \qquad (6.1)$$

Where $\tau_m$, $R_m$ and $E_L$ are the membrane time constant, resistance and resting potential respectively and $I$ is the input current. When $V > V_{thr}$ a spike is emitted and the potential is reset to $V = V_{reset}$. The correct implementation results in the behaviour recorded in figure 6.6, with the neuron's membrane potential rising to its threshold value (at -55 mV), whereupon a spike is emitted and the neuron is reset to its resting potential. The implementation was then altered to ignore the threshold checking (fig. 6.7a), leading to erroneous behaviour where the neuron does not fire (and consequently the membrane potential is not reset); the original code was then modified to introduce a more subtle error where the injection currents are caused to overflow the data-type used, leading to the evident misbehaviour of figure 6.7b.

Being able to visualise quantities such as these can greatly speed up the debugging and development process of new neural networks and component models.

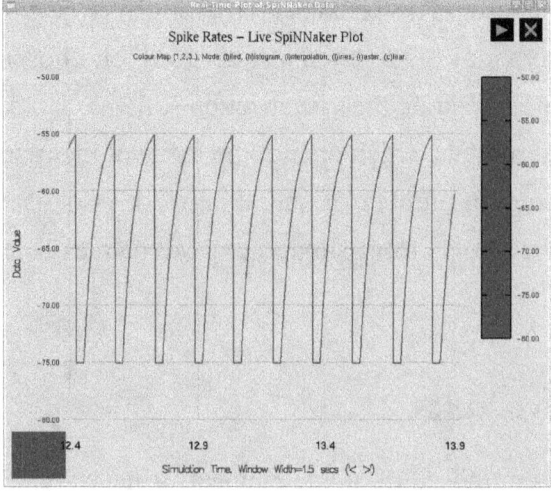

Figure 6.6: Correct neural dynamics for a LIF neuron injected with a bias current.

## 6.5. RESULTS

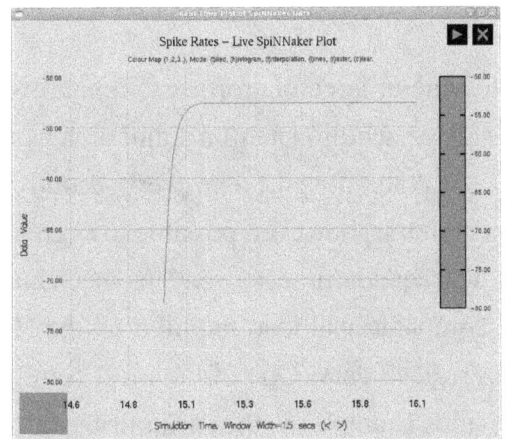 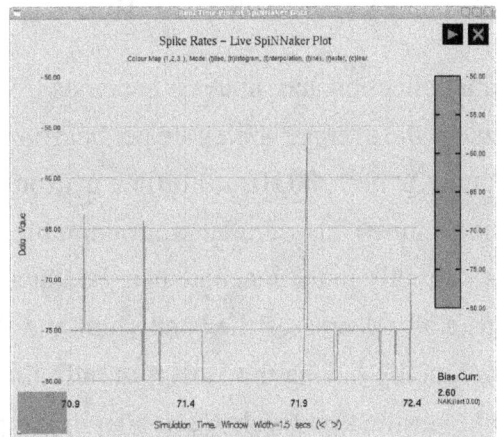

(a) Example of erroneous neural dynamics for a LIF neuron: threshold check not working

(b) Example of erroneous neural dynamics for a LIF neuron: overflow error

Figure 6.7: Visualisation of erroneous single-unit neuron dynamics.

### 6.5.2 Spike Activity

Spikes are the most basic observed activity of spiking neuronal networks and most commonly represented as raster plots. Here each neuron is assigned an ID on the vertical axis and time is represented on the horizontal; every spike for each neuron ID is noted at the time it arrives on the graph. The results are from a synfire chain model [VA05] which comprises 1,000 neurons subdivided in feed-forward connected pools; the signal propagation can be observed in the raster plot presented in figure 6.8, as well as a mean population firing-rate at the top of the figure, calculated at the visualiser by tallying the spikes received from this population.

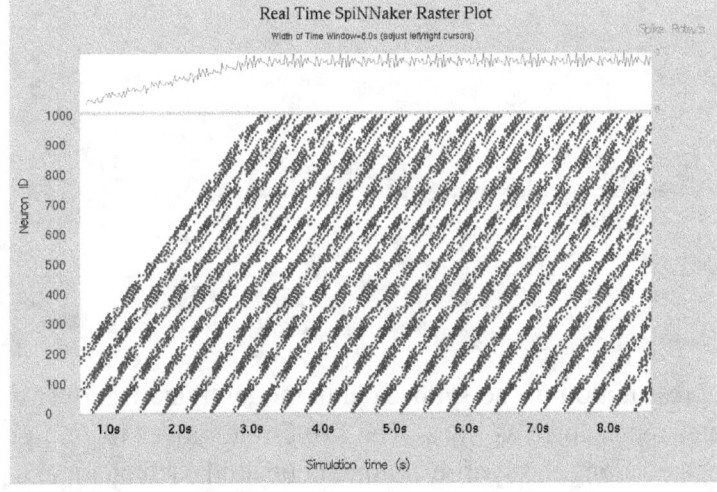

Figure 6.8: Raster plot representing signal propagation through a synfire chain.

### 6.5.3 Aggregated Information

Data collection and analysis becomes troublesome in large platforms; to report every spike if the average spike-rate per neuron was 30 Hz, a billion neuron simulation would require 32 bit × 30 Hz × 1 billion neurons = ~ 1 terabit of spike data every second. By using a massively-parallel programmable platform it is, however, possible to aggregate data directly in the machine using dedicated aggregation processors. This mechanism drastically decreases the bandwidth and the computational load required on the Host machine by having the Host plot only the aggregated data. One of the most common spike-abstraction levels (seen also in the previous example) is the mean firing-rate of a population of neurons, where the total number of spikes is divided by the number of neurons in the population. The collector node receives all spikes from a population, and aggregates this information before sending to the visualiser.

In this experiment a second synfire chain network was implemented where, rather than individual spikes, the in-system data aggregation is used to transmit only the population rates within the chain to the visualiser. The mean firing-rate $r$ of a population of neurons is calculated as:

$$r = \frac{n_{spikes}}{N \Delta T} \tag{6.2}$$

This is a measure of the mean activity in Hz across that population, where the total number of spikes recorded in the sample (sampled at interval $\Delta T$) is divided by the number of neurons in the population.

Figure 6.9 scrolls and plots the rate history of each *population* over time, and illustrates individual population spiking rates dying off as the stimulus is removed.

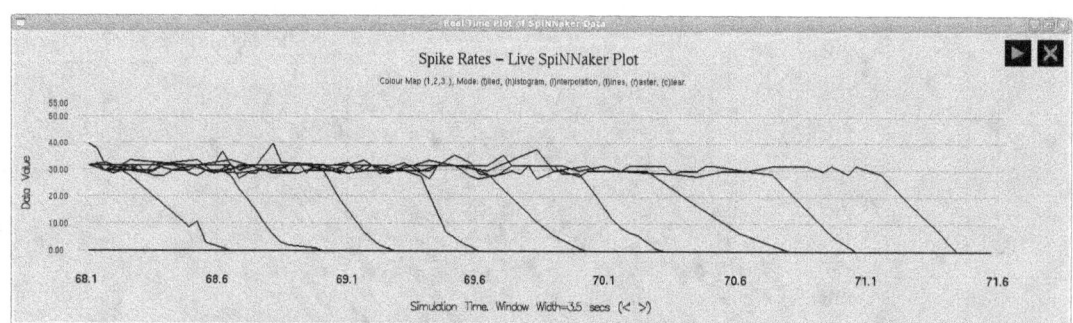

Figure 6.9: Representation of the synfire chain model dynamics over time with population firing-rates represented as lines. The firing-rate moves from high to zero as the signal is propagated along the synfire chain and beyond each channel's population.

## 6.5. RESULTS

### 6.5.4 Interaction

Whilst being able to visualise activity in real-time is good, being able to interact with a simulation is better. For this experiment a neural network simulation was created which can be adjusted based on user instruction from the visualiser. The population to interact with can be selected and adjustments can be made to the bias current injected into the population of neurons. The network created comprises 16 populations of 256 LIF neurons, each population assigned to a separate application core. The output is plotted in real-time to a user's screen so that the results may be monitored and activity displayed. To the left of figure 6.10 differing colours / tones represent the average spiking rates of the neurons in each of the 16 populations and the example population (3,1) was selected for interaction. To the right of figure 6.10 another real-time plot window was initiated from the GUI 'zooming' into this individual population to examine the firing pattern of its constituent neurons. This raster plot scrolls in real-time allowing the results to be observed as changes are made and, over the 10 second period of the plot, a reduction in the bias current was triggered, followed by a gradual increase.

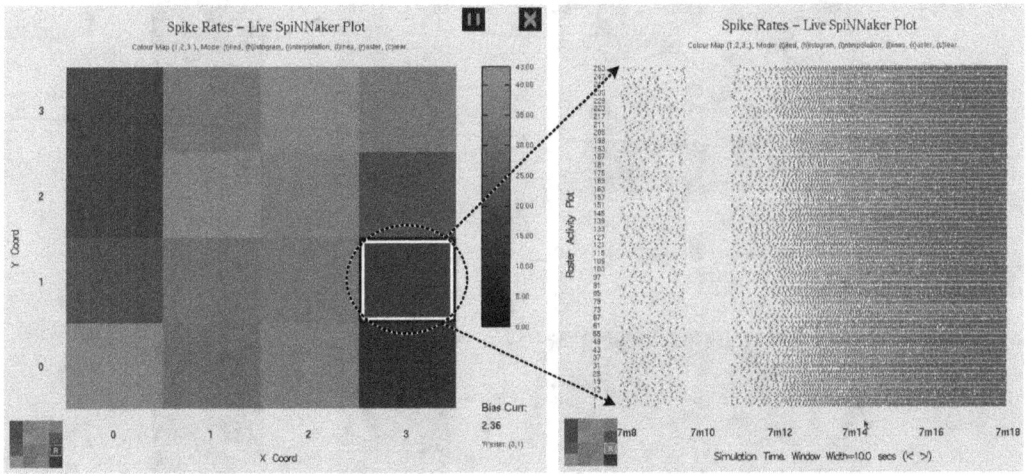

Figure 6.10: Interactive real-time neural network plots. Left: population rates (average spikes / neuron / s), users can adjust bias currents for their chosen population e.g. (3,1), and may open a second raster plot window (right) to view individual neuron firing events over time for that population.

### 6.5.5 V1 Modelling, Multi-View Simulation

As SpiNNaker is a general programmable modelling platform, it is possible to specify different network topologies and request different visualisation geometries. It is then possible to visualise activity from different populations directed to the corresponding sub-plot. An experiment was created to demonstrate this technique with a neural network which extracts the results of a bank of six Gabor filters from an image fed to the system through a web-cam. Its task is to reproduce the orientation selective receptive fields found in the primary visual cortex (V1) [DWH82]. A USB web-cam was connected to a standard GNU / Linux PC, which captured the image, sub-sampled it to a $16 \times 16$ matrix and fed this to the input neurons of the network as a bias current. Outputs from the network are a set of $10 \times 10$ matrices, corresponding to the output from each neural Gabor filter, computed as the weights between input neurons and the corresponding orientation filter. The output from the filters and camera is plotted on screen in real-time (fig. 6.11). Each output filter neuron fires at a rate proportional to how closely the detected and filter orientations match in that portion of the visual field.

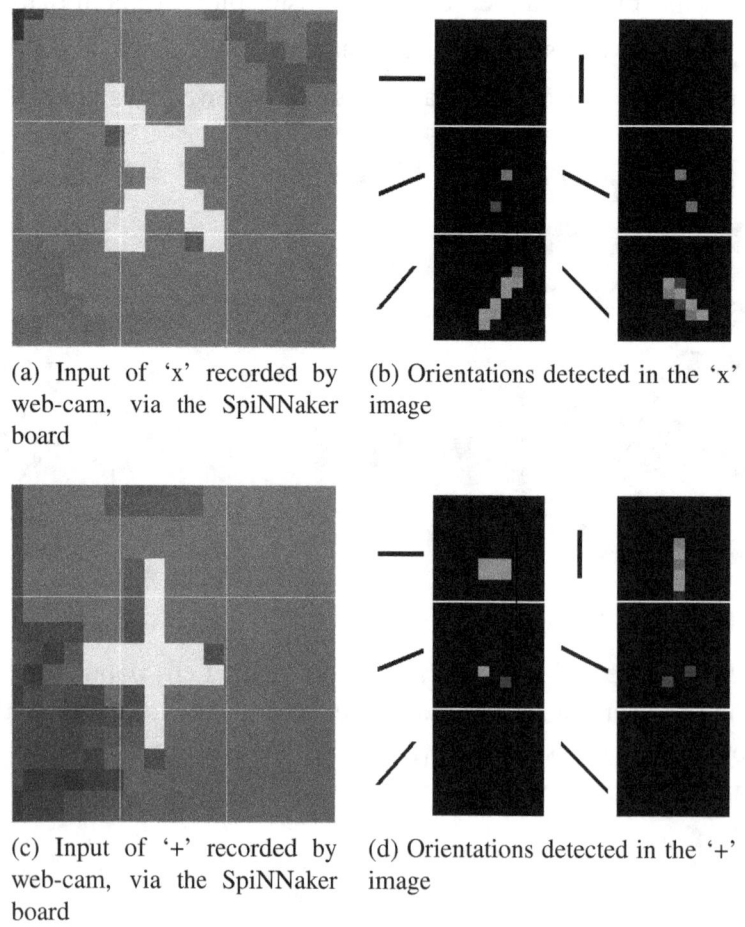

(a) Input of 'x' recorded by web-cam, via the SpiNNaker board

(b) Orientations detected in the 'x' image

(c) Input of '+' recorded by web-cam, via the SpiNNaker board

(d) Orientations detected in the '+' image

Figure 6.11: Screen captures from the output provided to the visualisation software in real-time from the SpiNNaker board with the V1 orientation ANNs in operation.

### 6.5.6 Manipulating Visualiser Data

Received data does not have to be plotted directly, manipulation can be performed, if required, in the visualiser module itself. In this final neural example the plot (fig. 6.12) is from the output of a network population (ensemble) created in the neural engineering framework (NEF [EA03]). This particular example is of a Cyclic Attractor [GDF[+]12] and the plot is formed by integrating the input at the visualiser over time, where the previous state (historic data) is used to calculate the latest value to plot. The output is oscillatory and to trigger this state the system is given a shock and in response the neural oscillator settles down to its characteristic frequency. This ensures that the immediate input and previous results are used to determine the latest data to plot. Due to the existence of the historical data it is possible to perform a range of manipulations and filtering as required within the visualiser itself – if this is not carried out at the aggregation point(s) within the network.

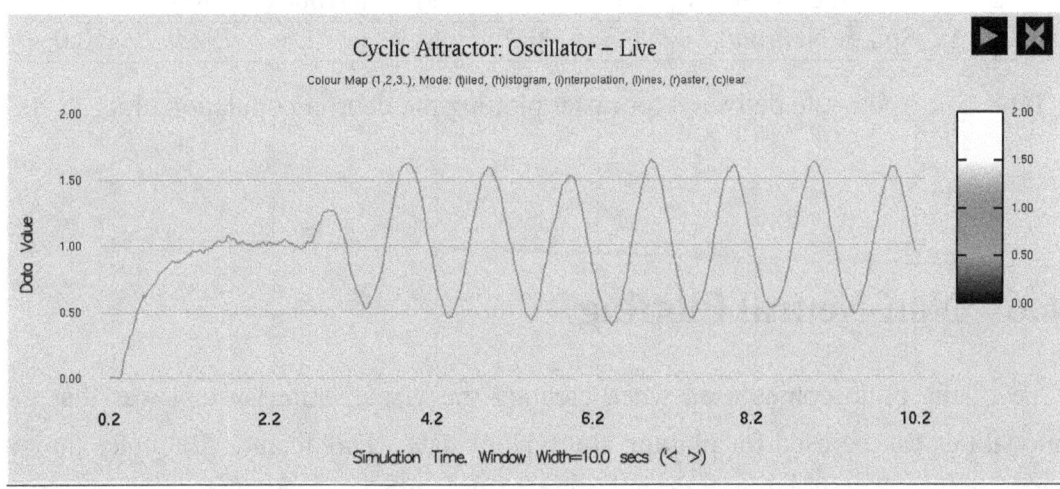

Figure 6.12: A Neural Engineering Framework (NEF) model running on the SpiNNaker platform may also output its data, an oscillator function is recovered via a first order filter on the output from a network.

### 6.5.7 Capacity of the Visualiser

To test the maximum performance of the visualiser with the SpiNNaker simulations the 'rate plot' simulation of section 6.5.4 was used (fig. 6.10). Its stimulus was incremented repeatedly (beyond that illustrated in the plot) and it was found that the number of plotted spikes plateaus at around 50,000 / s (see the highlighted entries on the right of table 6.1). This limiting factor was identified as the visualiser software and not the simulation or network connection of SpiNNaker. By simply reducing the frame rate of the visualisation software from 50 fps to 30 fps it was possible to plot all the spikes. Users must ensure that there is sufficient capacity in both target system and visualisation platform (and intervening networks) to support what is being requested. In this case, at 50,000 spikes / s the visualiser screen was solid blue, as per the latter stages of figure 6.10 would suggest. It appears a good case for filtering by aggregation techniques or rule enforcement to protect the simulation.

| Simulation Time | 7m8 | 7m10 | 7m14 | 7m18 | Plot Limit | Spike Max |
|---|---|---|---|---|---|---|
| Bias Applied (mA) | 0.75 | 0.60 | 3.12 | 5.26 | 19.5+ | 28.26+ |
| TX Spikes/Population/s | 1602 | 0 | 11575 | 18727 | 50944 | 64000 |
| Spikes Plotted (50 fps) | 1602 | 0 | 11494 | 18591 | *49371* | *49368* |
| RX SDP Ethernet frames/s | 843 | 0 | 978 | 1060 | 1463 | 1775 |
| RX Av. Spikes/Neuron/s | 6.3 | 0.0 | 44.9 | 72.6 | 192.9 | 250.0 |

Table 6.1: Spike-rate delivered for raster plotting the neuron population of fig. 6.10.

## 6.6 Non-Neural Plotting

One of the philosophies used when creating the visualisation system was that the modalities can be used for plotting any type of data. This feature decouples the visualiser from neural network visualisation on the SpiNNaker platform and permits it to be used as a more generalised tool.

**Heat-Map** SpiNNaker, as a general purpose programmable platform, may be used for parallel programming tasks more diverse than just neural networks. One such example is the 'Heatmap' demonstration of parallel processing and message passing.

This simulation model uses 64 Application Processors of a SpiNNaker 4-chip test system to calculate the temperature at $8 \times 8$ points on a rectangular surface – which has stimulating temperatures applied to all four edges. Once a change in temperature is applied interactively by the user, the system proceeds to iterate in parallel across the simulated material using its thermal characteristics as a model, passing messages to neighbouring cores (points) until a thermal equilibrium is reached. This example

## 6.6. NON-NEURAL PLOTTING

fits nicely onto the tiled visualisation modality, and a visualisation decoder module was created to parse the data arriving from the simulation. For the interaction section (see the bottom right of the visualisation), a custom block was written to display the temperature stimuli, and when the user clicks into the relevant box and adjusts the temperature with the standard + / - or mouse scroll wheel, these inputs are sent to the simulation platform for processing.

Three figures have been included by means of an illustration. Firstly figure 6.13a is a tiled plot of the initial state, from which further experimentation is performed. The second plot (fig. 6.13b) is an interpolated version of the experiment where the values have been adjusted via the user controls in the lower right. The third and final illustration (fig. 6.14) uses the 'line' function of the visualiser, where changes in all 64 data values are plotted over a 30-second period. The lines display the series response of each data point to the changing temperatures over time. The colour scheme may be changed at any point via the menu or a keyboard shortcut, and the ribbon key on the right-hand-side illustrates how the temperature data is represented.

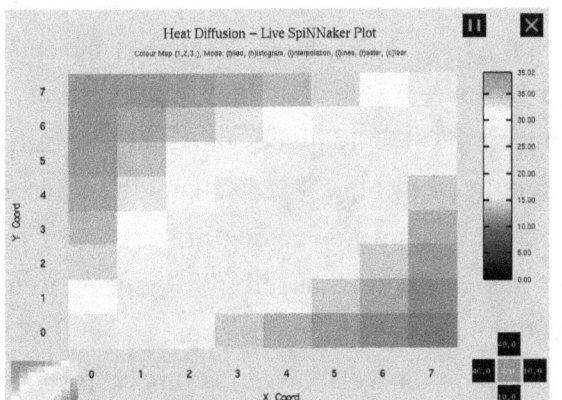
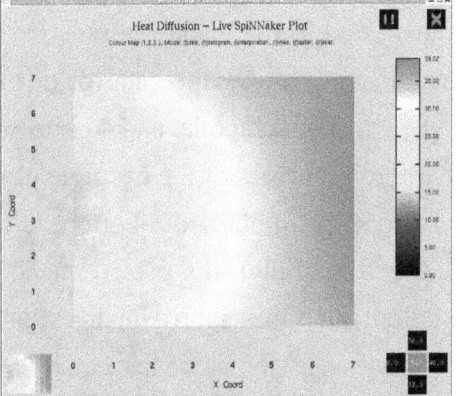
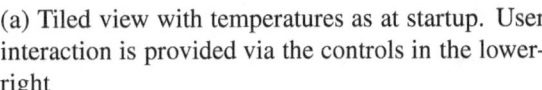

(a) Tiled view with temperatures as at startup. User interaction is provided via the controls in the lower-right

(b) Interpolated view of altered temperature plot. The small tiled version in the bottom left shows the original data

Figure 6.13: Real-time visualisation of the non-neural heat-map application executing.

**Processor Utilisation** As the visualisation platform is versatile it may be deployed to plot any set of data including hardware status. In this example a real-time visualisation of the utilisation of the individual application cores on multiple SpiNNaker chips has been created. The application cores are executing a synthetic application which generates a load proportional to the identification of the core, so core 16 for example has

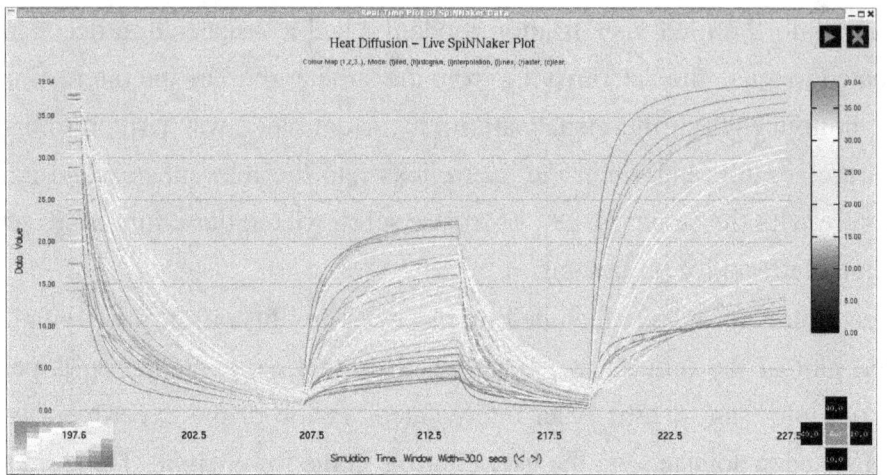

Figure 6.14: Line diagram of how the individual channel temperatures change over time based on the thermal characteristics of the simulated material. This example shows the transition from fig 6.13b to all-zero, then an adjustment to a randomised set of values, back to zero and reset back to the starting state.

four times the work to do as core 4. The status of all processors on each chip is transmitted to, and consolidated by, an aggregation processor in chip (0,0) before being transmitted to the visualisation workstation. The visualiser has a small visualisation decoder to interpret the data for display. In figure 6.15a all 4 chips are operating the same coreID:load code, and 4 chips with their $4 \times 4 = 16$ processors can be identified in the four corners of the example plot. Within each chip's $4 \times 4$ subplot the cores are plotted sequentially from bottom left in columns upwards ending in the upper right. In the second plot (fig. 6.15b) the colour map has been changed from 'greens' to use a colour map that approximates what would be returned by a 'thermal camera'. The option to view the value has been turned on and this is displayed in numeric form and this updates in real-time.

Calculation of the percentage utilisation is carried out on board the aggregation core. At any particular moment the active / asleep status of a SpiNNaker core may be polled from the system controller. A history of this binary status is stored and a weighted average calculated over the previous second for transmission to the visualiser. The polling frequency for this dynamic information is important, as if it is co-incident with the period of the workload, a false aliased representation and / or beating may occur. Two techniques may be used to cancel out this problem, the first of which is to poll more frequently than twice the periodicity of the minimum workload (Nyquist's criterion), the second is to introduce randomised jitter into the sampling process to avoid the problems of synchronisation when averaged out over a long enough period.

## 6.7. SUMMARY AND CONTRIBUTIONS

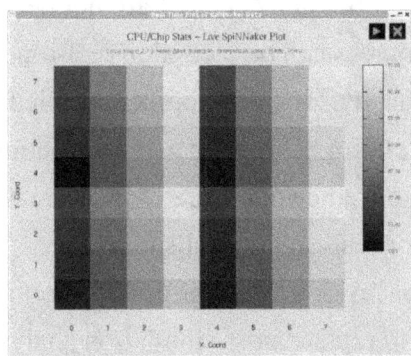 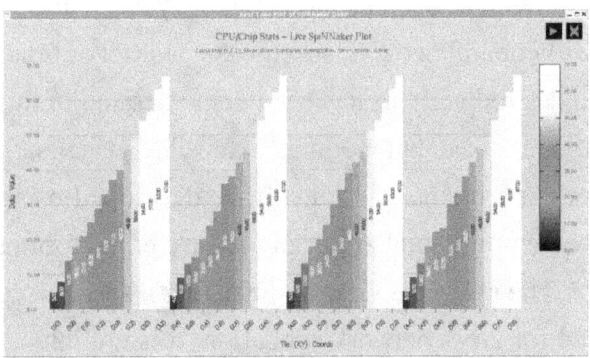

(a) Utilisation – by chip represented tiled per position on the board

(b) Utilisation – as a 1D histogram, but including numeric labels on the data

Figure 6.15: Real-time visualisation of the processor utilisation of SpiNNaker cores.

## 6.7 Summary and Contributions

As models of neural networks scale in concert with increasing computational performance, gaining insight into their operation becomes increasingly important – especially as they pass into the realms of biological significance and operate at or near real-time. This chapter has proposed and demonstrated a generalised method to visualise data at all layers of SpiNNaker neural networks, providing representation of the network in many forms in real-time. By providing flexible access to all layers and elements within a neural network simulation as it runs – from neurons and synapses to hierarchies of populations, and the network as a whole, an interactive and more easily understood picture of the network emerges and can be presented to the users.

The key contribution from this work is the management visualisation tool itself which enables users to gain one or more viewpoints on the performance of their simulations in real-time. Conventionally the determination of network correctness occurs post-simulation, but with sufficient 'in-flight' insight from this tool, malfunctioning ANN simulations may be terminated early to save computational resource, or the user may use interaction mechanisms to intervene (or automate remediation). Longer-term simulations will benefit from real-time evaluative facilities and, as SpiNNaker targets the real-time space, then longer-term simulations are expected as the interactive loop is closed. Conversely, at the small and short-term scale, visualisation is also important when designing and prototyping small-scale networks to verify correctness.

In terms of novelty and furthering the field, this work recognises that many of the most common neural network plots are generalisable and modalities may be applied to different parameters and network-types. Furthermore, the plots created are not necessarily neural network specific, so a variety of data sources can be plotted in real-time if so desired and this has been demonstrated in this chapter. The development has led to a generalised, but modular, extensible visualisation platform, where new visualisations for experiments can quickly be developed using the framework already in place and where non-neural applications have equal footing for real-time visualisation.

The real-time view of the data is a specialised space – it is used for troubleshooting and for gross network validation. It is not intended to replace post-simulation analysis where diverse analysis tools can be used and more accurate temporal data resolutions are available (as the network over which the real-time visualisation data flows is non-deterministic). In this respect it is intended to provide a similar rôle to that of functional imaging of the brain, where a focused viewpoint is taken but does not examine every cell simultaneously.

Although this chapter has commonly referred to real-time network simulation, it may not be possible to achieve this time-basis in all cases – some may run faster, and some slower depending on the size and complexity of the network and its components. This does not preclude use of the visualiser or its interactive features as it is possible to interact at whatever time-base is used by the simulation.

In the development of the visualiser techniques for requesting and switching data, aggregation within the SpiNNaker system is used, thus allowing zooming to be performed efficiently by focusing only on those areas which are of interest to the experimenter. This aggregation portion is also a key contribution as, without it, it would prove impossible to scale the visualisation system to larger networks and viewpoints. The aggregation techniques also perform their rôle without a significant burden being placed on the software and hardware of the SpiNNaker platform, taking advantage of the inherent parallelism available in the system being monitored itself.

# Chapter 7

# Managing Large Network Attached Systems

Managing the availability of computing facilities and infrastructure is a discipline known as Systems Management. This field concentrates on the monitoring of physical equipment (and perhaps the services which run upon it), to locate faults reactively and begin the remediation process as quickly as possible thus minimising service disruption. In addition to fault management, systems management often covers a range of proactive rôles including auditing, security and capacity planning.

While systems management is an operations centric view, alternative analytical systems exist which take a different perspective. From a business activity viewpoint for example, if a server goes down the sales team cannot process any orders, and the revenue stream is cut. This top down approach to monitoring may be more appropriate to the intended audience than a bottom-up systems approach.

In this chapter, and in the context of managing large network attached systems, systems management of an operational flavour is examined.

## 7.1 Principles of System Management

Perhaps the best known and well defined model of systems management best practice is the 1989 framework: ISO Standard ISO / IEC 7498-4:1989 [ISO89b]. This model identifies requirements such as system reliability, predictability and the necessity that users must be able to gather information and exercise control via tools in the management environment. The framework identifies five main management functional areas: Fault, Configuration, Accounting, Performance and Security (FCAPS).

The ISO originally created model specifications in each area with the intention that tools and protocols would be created to cover each separately. However, as many of the management areas overlap practically, management tools do not apply the distinctions rigorously, and the ISO working group decided to merge the protocol specification for all five areas into Common Management Information Protocol (CMIP) [ISO89a]. One key principle remained: the decoupling of the structure of the management data (known as the MIB (Management Information Base) [MR88a]), from the protocol used to access it. Although CMIP was never a great success, the alternative protocol Simple Network Management Protocol (SNMP) [CFSD89] was, also maintaining this principle of protocol and data separation. SNMP found favour with equipment and software vendors who used it as their systems / network management protocol of choice.

The ISO FCAPS model is applied practically in systems management today, for instance, underlying Cisco Systems' 'Network Management System: Best Practices White Paper' [Cis07]. The sections below provide a brief overview of the five main management functional categories:

### 7.1.1 Fault Management

This functional area covers detecting faults in the system and isolating or correcting them as swiftly as possible. The management tool used for the fault management function should handle notification messages from the system (including level of importance), and be able to initiate diagnostic testing and provide the opportunity to implement remedial measures. Examples of such system management tools include: IBM Tivoli Netview [IBM12c], CiscoWorks [Cis12a], HP Openview [Hew12c], and Nagios [NE12]. These fault management tools are populated with device data and are able to poll proactively or reactively receive event notifications for handling. Actions that may be triggered include updating a status screen / dashboard, sending emails and texts, automatically raising a problem ticket, sounding an alarm or other method of notifying the appropriate user(s). Although these event management tools primarily alert operators to problems, in some circumstances they may initiate automated corrective actions, such as restarting a service, or rebooting a piece of equipment.

### 7.1.2 Configuration Management

Configuration management gathers configuration data and exercises control over a managed system, whether that configuration is in planning, operationally deployed, or out of service. Areas covered include configuration files (and versioning of such documents), carrying out of software and hardware inventories, and collecting information on the condition of the system configuration. Examples of tools performing configuration management are version control tools such as Concurrent Versions System (CVS) [Fre06] which logs and tracks changes to source files in managed software development projects, Dell's hardware appliance Kace [Del12b] which automates software upgrades for consistency of support, and CFEngine [CFE12] a software compliance checking tool which may be used to validate a business's infrastructure against regulatory requirements.

### 7.1.3 Accounting Management

The account management function keeps a record of resource utilisation, with the ultimate goal of being able to identify and categorise utilisation of a system. This information may be used to initiate billing, to trigger investigations into abnormal use, or perform statistical analysis or capacity management activities. Examples of where such management facilities are used include telecommunications billing, or in high performance computing where processing time is booked, logged and charged back to users (for example NASA's High-End Computing Program [NAS11], or the Open University High Performance Computing Cluster [Uni11]).

### 7.1.4 Performance Management

Performance management's goal is to examine the behaviour of the managed system resources, to understand and evaluate current operating statistics and to enable its subsequent analysis. This information is also used to carry out trending and capacity planning, ensuring that performance of the managed system remains within acceptable limits. Examples of performance management software are the open-source tools Cacti [Cac12] and MRTG [Oet11], which gather data on resource utilisation to present as graphs or reports and assist in mid-to-long term capacity planning. Other examples include the Unix 'top' command or the Windows Task Manager to determine resource utilisation of processor or memory allocation on a machine. The output from a performance management tool can be used to trigger actions based on a threshold, for example to terminate a process, bring more capacity on-line (or take excess capacity off-line!).

### 7.1.5 Security Management

The purpose of security management is to allow application of a defined security policy to the management system, and may include user access control, encryption of sensitive data, and blocking unauthorised remote access. Security management of a system should be implemented as part of a wider security policy that includes physical access control and user validation. Security management also covers the logging of data such as authentication and authorisations, to ensure that it is possible to reconstruct events and fully audit the system against the applicable security policies. An example security management function is access control of remote connections into a company or institution (e.g. a VPN, or SSH console access).

## 7.2 Managing Large Systems

System Management is an important suite of functions ensuring the health of complex computing systems and services. The majority of large-scale systems are connected via networks, and therefore may be located where suitable environmental facilities can be provided. Monitoring of such systems is therefore carried out remotely.

### 7.2.1 Remote Monitoring

Remote monitoring is now ubiquitous in systems management applications with the inevitable reduction of headcount, improved scalability and reliability this affords. This technique has become convenient due to the reduced commodity pricing of networking connections and may be enabled by two separate paths:

**In-Band**

In-band implementations use the same network path for both application and management traffic. An example of this option is gaining access to the control interface of a web server via the same Ethernet connection that it uses to serve up web-pages to end-users. This clearly has the advantage of simplicity and economy, but the disadvantage that the command path to the server is vulnerable to failures or attacks on this single connection. Additionally excessive amounts of either management or application traffic may compromise the other class.

## Out-Of-Band

With Out-of-Band (OOB) signalling, a separate path is provided for command and control purposes. Extending the earlier example, a separate line and modem may be connected to the server, adding complexity and cost, but providing diversity from the data-plane. Appropriate security measures must be taken for all OOB connections that are available via a public telephony network or IP address.

## Selecting In- / Out-Of-Band Management

An important decision taken in the design of any large system is whether management information uses either in- or out-of-band signalling paths. This choice may involve a number of factors:

1. Cost – extra equipment and telecommunications charges are usually incurred when choosing OOB management, can the costs be justified?
2. Complexity – how much extra equipment (and support) is required to provide the OOB access (and can it fit in the space available)?
3. Downtime – how critical is it for the managed service to be 'up' – what Service Level Agreement (SLA) has been set or what proportion of downtime can be tolerated?
4. Proximity – is the equipment in a machine hall in the same building, or in a remote location to support staff – how much does it cost, and how long does it take to get someone on site?

### 7.2.2 Hardware Management

While it is possible to provide system management in-band, this solution tends to be used for non-critical and cost-conscious applications. Here the management traffic shares resources with the data plane, but mechanisms such as quality of service can be applied to traffic to aid in prioritisation of network or computation resources. In the realm of performance computing, in-band management is an atypical choice, particularly in critical applications, where separate OOB connections are usual.

**Intelligent Platform Management Interface (IPMI)** The majority of Intel-based server manufacturers incorporate IPMI compatible facilities into their hardware. IPMI [Int09] is a message-based mechanism that permits system operators to independently monitor and manage the physical properties of a system. It uses its own Baseboard Management Controller (BMC) that is independent of the in-band processing path, giving access to instrumentation and providing control options for the system. This operates regardless of the main server status and enables remedial actions to be invoked remotely, such as power-cycling a non-responsive server. Each manufacturer is free to adapt and extend their IPMI implementation and the major server vendors have their own solutions tied to their hardware, such as HP's Integrated Lights-Out (iLO) [Hew12b], Integrated Dell Remote Access Controller (iDRAC) [Del12a], or IBM's Remote Supervisor Adapter (RSA) [IBM12b].

Two techniques are used in IPMI to isolate management traffic from the data plane. More usually a full out-of-band connection provides a totally distinct path to an external connection (such as serial or Ethernet), or alternatively a side-band management path may be provided which takes a VLAN 'tap' from the main network connection for the server, which is diverse from that passed to the main board. These OOB management paths may then be consolidated into a separate 'management' network, and IPMI may be extended to provide full remote-access to the keyboard, video and mouse (KVM) connections. The KVM Input / Output connections from the server main board can be made available via the IPMI management path, giving full remote access to items such as the BIOS, which would normally require an on-site presence. IPMI implementations are generally able to be daisy-chained with same-vendor equipment, and may not require all devices to be wholly homogeneous, which is ideal for server farm type deployments.

**Cluster** The above IPMI systems management model can also be used when many of these systems are interconnected to form a cluster, providing a high-performance computing resource by aggregating the processing capability of the constituent machines. In many of these instances applications may be communications-bound so separate high-performance (but standard) interconnection networks such as Infiniband [IBT10] may optionally be provisioned for message passing, intercommunication and system purposes. The server vendors are once again strong in the cluster management environment, as equipment in this configuration tends to be homogeneous, for example HP's Insight Cluster Management Utility [Hew12a] and IBM Cluster Systems Management [IBM12a].

**Grid Computing** Grid resource management is necessarily more ad-hoc as not all the constituent computation resource may be dedicated, or indeed in the same management domain. Systems management of grid systems therefore typically targets the resource available to perform work, and relies on the constituent computing nodes registering with a resource management system (RMS), or the RMS having a pre-populated list of machines [KBM02]. Tools in this space are typically less commercially oriented – as there may be no particular vendor relationship – and open-source solutions are commonly used (e.g. Netlogger [GTC+00] for performance analysis, and Condor [LLM88] for workload management). In the event of a soft or hard problem with a machine in the grid, the job will require reissuing dynamically to another area of the system, and the RMS is left to take care of the reallocation process.

**'Supercomputers'** In HPC computing there is significant focus on attaining best utilisation of capacity, and this drives much of the management software available, which has a heavy focus on job scheduling [THW02, HKKS03]. For system monitoring of very large HPC solutions (supercomputers) a range of vendor provided, open-source, hybrid and home-grown techniques are employed, as outlined in a 2011 discussion paper on HPC monitoring [BAFL+11].

For example on IBM Blue Gene HPCs, Service and Input / Output nodes maintain database information about the system state [ABB+03]. This information is made available in the Blue Gene Navigator, but IBM recognise that a mix of management techniques may be used, and have published 'Red Book' guidelines on how popular system monitoring and management tools such as Nagios can be deployed alongside their own tools [DHMW08].

**A Distributed System – The Internet**

Of course not all systems to be managed are HPCs. A massive distributed system is found embodied in the switching and routing infrastructure equipment of the Internet. Its hardware systems are deployed under the management of many competing, but collaborative domains of command and control. SNMP is almost universally used in these interconnecting networks [KWC+09], providing device-centric information on deployed hardware, and statistical information on data-flows amongst them.

### 7.2.3 Systems Management Software

Management software is often vendor specific – manufacturers provide software tailored to their particular hardware (or software) solution. This proprietary approach, however, provides little flexibility when considering monitoring many diverse systems concurrently, particularly where multiple vendor's equipment is used. A standardised, consolidated approach to system management is often used in such environments and where common management frameworks and protocols can be exploited.

There are a large number of commercial software packages available for generalised systems monitoring: example providers include PRTG [Pae12], Big Brother Software [Que12] and SolarWinds [Sol12]. There are additionally many open-source alternatives, SpiceWorks [Spi12], GroundWork [Gro12], OpenNMS [Ope12] and of particular note – Nagios [NE12]. Nagios is widely and diversely deployed, providing monitoring capabilities across a number of device-types and applications. In addition to it being 'free', Nagios has the flexibility to support multiple protocols and is readily extensible via scripting – all of which have helped nurture its popularity.

IBM have recently augmented their presence in the generalised systems management space with their acquisition of vendor-agnostic Platform Computing [Pla12], who produce a portfolio of technical system management tools covering a range of high performance / throughput platforms, with grid, cluster and supercomputing support.

### 7.2.4 Protocol and Schema Standards

To understand the health of the system being managed, a monitoring system must be put into place. There have been several attempts at providing standardised methods of retrieving management information from systems. A number of protocols and strategies have been defined to cover both network and systems management, and this section covers a number of the more significant approaches in this area.

**WBEM** One such initiative to unify the management of computer systems, particularly those that are distributed, is Web-based Enterprise Management (WBEM) [DMTF10]. WBEM defines a set of standardised techniques which together provide a common method for management data to be gathered and exchanged between managed resources. WBEM defines protocols, mappings and discovery mechanisms, with the Common Information Model (CIM) [Dis12] standard being used as the basis of hardware and software object representations.

## 7.2. MANAGING LARGE SYSTEMS

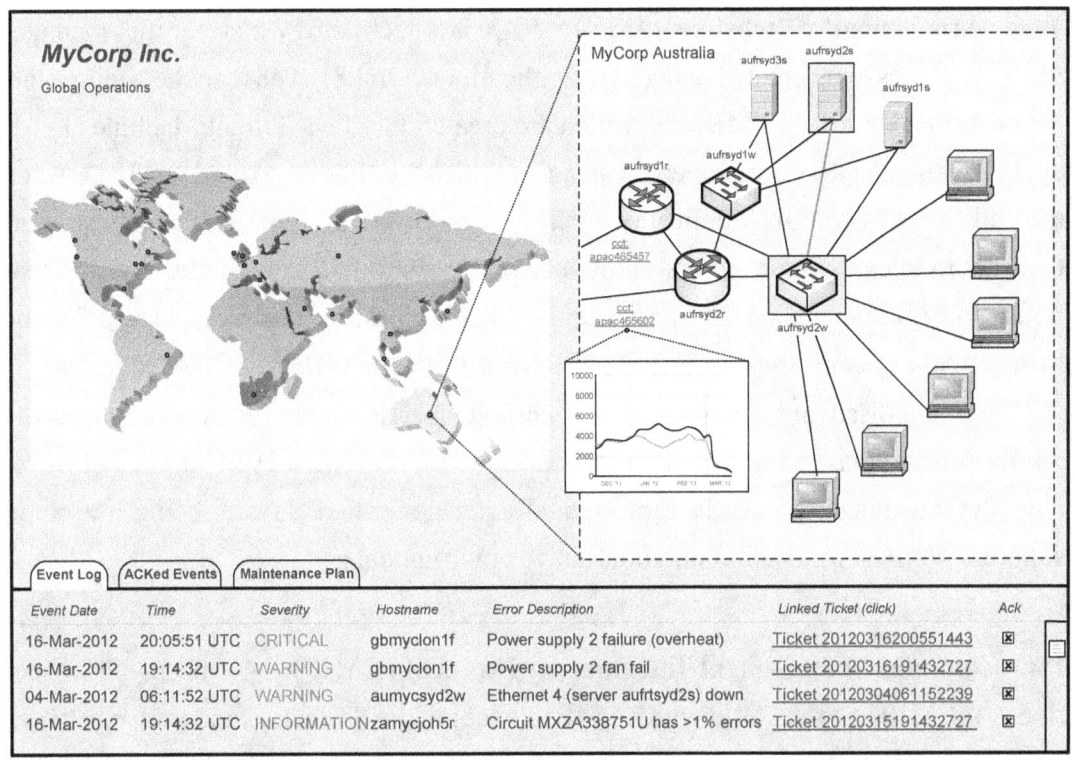

Figure 7.1: Typical management software view. A console presents events by severity and chronologically, and a graphical interface presents the real-time status of items managed topologically, which may then be 'zoomed into' to explore in greater detail.

CIM provides a consistent vendor independent method of representing managed objects and their relationships, whilst its syntax remains extensible to include vendor specific information. To transport data between managers and entities the standard HTTP(S) protocol is used, with information encoded using XML objects.

Windows Management Instrumentation (WMI) is the Microsoft implementation of WBEM / CIM, and is available on its operating systems. It provides access to managed objects using a unified data model, which integrates with the Windows programming environments, event handling and scripting techniques. WMI is perhaps the most widely deployed implementation of WBEM, but is limited in its application to the Windows family of Operating Systems. In addition to the WMI implementation, many GNU / Linux distributions include open-source WBEM implementations including OpenPegasus [OG11] and OpenWBEM [Ope06].

Once WBEM components are combined with the objects being supported it may be used as an end-to-end (remote) management solution. WBEM is able to represent relationships between managed objects, and standard protocols are used across the board, however it is by no means universally deployed across infrastructure equipment.

**Java Management eXtensions (JMX)**   JMX is a technology that permits management interfaces to be created using Java applications [Ora08]. The managed end-points (or probes) are known as MBeans, which are created in a Java Virtual Machine (JVM) environment, and they are probed by an agent which straddles the relationship between the end MBeans, and the applications which monitor the system. The agents support two methodologies, 'connectors' allow native access to MBeans by the end applications, and 'protocol adaptors' translate the MBean into the desired protocol for the management system, such as SNMP or HTTP / XML for WBEM / CIM techniques.

The main disadvantage of the JMX technique is that it requires Java to run, so the platform which has the items to manage needs to be running a JVM. It does however have the advantage of easy integration and extension where Java is being used, and supports protocol translation into non-native environments.

**ISO Common Management Information Protocol (CMIP)**   As discussed in section 7.1, CMIP lost the battle with SNMP to become the de-facto standard for network management protocols. The take-up of CMIP was low primarily due to its potential to overwork the resources on the monitored device, and that the alternative SNMP was lightweight in comparison and thus less costly to implement. This together with the proliferation of TCP/IP which is the native SNMP protocol (CMIP was originally designed to operate on the OSI protocol suite which was not a success either), resulted in the CMIP management protocol falling behind in its use, and all but dying-out for widespread and commercial use [Dov12].

**Simple Network Management Protocol (SNMP)**   Whereas CMIP failed to garner support, SNMP was successful, but maintains the common philosophy of having a distinct protocol for transporting the data, and a separate Management Information Base (MIB) definition which includes the structure of the objects to be managed. SNMP is the most prevalent management protocol in use today, with bespoke MIB structures created to support a vast array of equipment and software.

SNMP has drawbacks, mainly in areas of security and efficiency as the overhead on each (unencrypted) request is substantial and there is no native aggregation function. The continued success of SNMP self-perpetuates into near ubiquity of support from hardware system manufacturers, particularly as it is relatively simple for new management trees to be created as required, for whatever application, whether it be hardware or software.

**Consolidation of System Management**  It is possible to create a bespoke system management solution, but with larger and more complex systems, it often makes sense to choose one of the existing management frameworks. Systems management is also about providing a consolidated view of the pertinent information for the users. Most systems managers would prefer all alerts to be presented by a single system, and here the more general system management tools have an advantage over the niche vendor-specific tools. A common component in the vast majority of systems management software is the support of SNMP to monitor devices and equipment remotely, with most network attached equipment having management MIBs provided for them for exactly this purpose. SNMP's ubiquity, and the MIB's flexibility drive their continued use for management purposes. This chapter therefore continues with a review of the SNMP management framework.

### 7.2.5 SNMP: A Walk Through

Simple Network Management Protocol (SNMP) was proposed in Request For Comment RFC1067 [CFSD89], and designed to be a simple low overhead protocol to allow Network Management Systems (NMS) to set and retrieve information from multiple network attached elements. It was produced by the Internet Engineering Task Force (IETF) in response to the Internet Activities Board (IAB) requirements for a management solution for network attached devices [Cer88], and at around the same time as the ISO CMIP alternative. The management information to be accessed by the NMS devices is stored on the to-be-managed systems (agents), in the tree-like Management Information Base (MIB) structure [Pre02]. The MIB is formed of specific objects that the agent maintains, such as statistical counters and gauges. As well as being located on the agent, the MIB may supplied by vendors in a standard format to be compiled onto any NMS platform that needs to use it. Multiple NMS systems can be attached to the network to attain resilience, or alternative viewpoints for diverse audiences. In larger systems a hierarchy of NMS machines can service a network of devices, and agents themselves can also be organised into their own hierarchy, with master / slave agents to help partition the management tasks (AgentX [DWEF00]).

SNMP became popular quickly [Kas91], and has remained so due to its simplicity and extensibility, its being free to use and its mature toolsets and vendor independence. SNMP agents and implementations can be found in most types of device that can be network attached, including servers, networking equipment such as routers and switches, and end stations such as PCs, workstations and printers [Sim12].

154     CHAPTER 7.  MANAGING LARGE NETWORK ATTACHED SYSTEMS

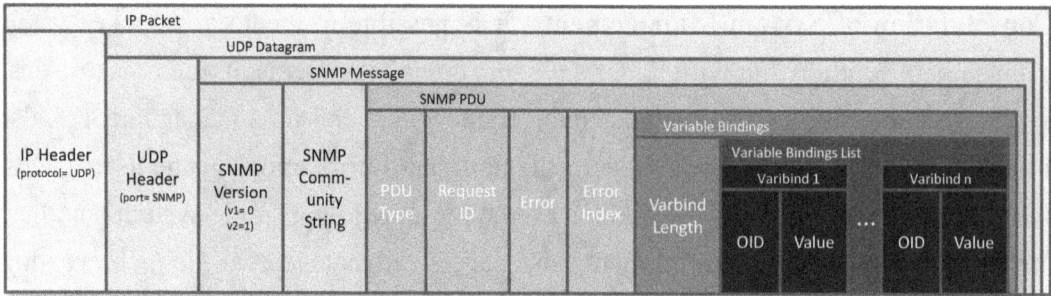

Figure 7.2: IP packet encapsulation of SNMP (version 2c).

**SNMP Protocol and Messages**

Internet Protocol (IP) is predominantly used to convey SNMP messages between managed agents and the Network Management Systems (NMS). There are three categories of SNMP message:

- Get – this is the NMS requesting data from the agent, and is responded to with a get-response. Request variants include, get, get-next, get-bulk.
- Set – this allows an NMS to set a writable value in the MIB, and the agent replies with a set-response.
- Agent Initiated – this includes messages such as 'trap' which inform the NMS of notable events / alarms on the agent.

Get and set messages use UDP port 161 for SNMP transport, and UDP port 162 is used for the agent initiated traffic. An example SNMP packet structure with IP encapsulation is illustrated in Figure 7.2, and as shown it is possible to convey more than 1 SNMP request / response (variable bindings) within the same packet.

An example 'get-response' message is captured from the network in Figure 7.3, formed of the initial 2 field message header (containing the version number and (in version 1 and 2) the community string which is used as a 'shared secret' for authentication). All SNMP data is transmitted without encryption between managed agents and the NMS, until SNMP v3 where authentication and integrity checking is added, however this version is not as widely deployed as the most common SNMP v2c form. The header is followed by the SNMP PDU (Protocol Data Unit) which is formed of:

- Type – the example in Figure 7.3 is a get-response
- Request ID – to match up specific responses and requests
- Error Status – null, otherwise indicates type, then Error Index indicates the object ID which is in error

## 7.2. MANAGING LARGE SYSTEMS

- Variable Bindings – including Object Identifiers (OIDs) and data, there may be more than 1 per message to improve the payload:header efficiency

```
▽ Simple Network Management Protocol
    version: v2c (1)
    community: cameron-thesis
  ▽ data: get-response (2)
    ▽ get-response
        request-id: 570271233
        error-status: noError (0)
        error-index: 0
      ▽ variable-bindings: 1 item
        ▽ SNMPv2-MIB::sysName.0 (1.3.6.1.2.1.1.5.0): camerons-device.cameron.patterson
          ▷ Object Name: 1.3.6.1.2.1.1.5.0 (SNMPv2-MIB::sysName.0)
            SNMPv2-MIB::sysName: camerons-device.cameron.patterson
```

Figure 7.3: Packet capture of a response to an SNMP get operation, returning data from the agent to the NMS.

### Remote Monitoring (RMON)

Remote MONitoring (RMON) [Wal91], is a protocol used to partially distribute the collecting and analysis task to devices within the network. It runs as a 'probe' on remote devices and is able to collect detailed statistics about what is occurring locally as the probe is proximate to the monitored device. RMON Probes reduce the quantity of data transmitted by aggregating data and sending it back to the management station only as required. Usual SNMP get / set / response packets have small payloads, and therefore a high percentage of the data transmitted is overheads and encapsulation, therefore collating and aggregating data locally before transmission is more efficient.

### 7.2.6 The Management Information Base (MIB)

**MIB Construction**

The MIB is a structured tree of management information, comprising data objects that may be retrieved (and / or set) by management systems on demand. It may also define agent-initiated events / alarms / notifications that are sent from agents to NMS when a pre-defined set of circumstances occur. The MIB specification which is most commonly used today is based on MIB-II [Ros90] and successors. It is independent of the protocol which is used to convey it over the network.

156  CHAPTER 7. MANAGING LARGE NETWORK ATTACHED SYSTEMS

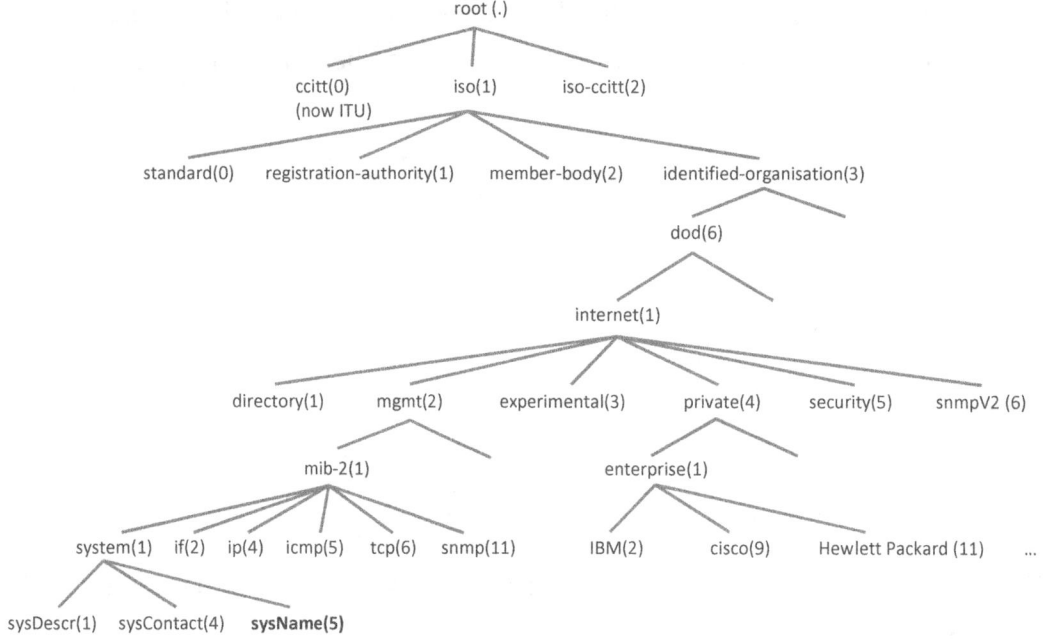

Figure 7.4: MIB tree structure, including system name, and links to the root of the private.enterprise MIB for several vendors.

The object tree is generally fixed into the agents' operating software, with manufacturer and product specific information grafted onto the private.enterprise part of the MIB tree (fig. 7.4). So that generic NMS software can understand individual enterprise MIBs, they may be supplied by the manufacturer and loaded / compiled into the NMS software, or items added by hand from the MIB on an ad-hoc basis.

MIB objects are guaranteed to be globally unique, such that each standards body or private enterprise is assigned its own root identifier in the tree, under which they define their objects to be managed. MIB objects (stored as the leaves of the tree), are referred to by their OID (Object IDentifier) which is a dotted decimal notation, but there is also a textual representation used to aid human comprehension of the MIB and the information it describes. An example of both formats used to retrieve the hostname of a system follows: '*.iso.org.dod.internet.mgmt.mib-2.system.sysName*' proves a more readable version of the equivalent numeric: *.1.3.6.1.2.1.1.5*. A view of this particular object in an example MIB tree has been highlighted in the lower-left of figure 7.4.

**Format of Data Storage**

Part of the IAB requirements in RFC1065 [MR88b] is for structuring the format of data stored in the MIB tree, including such items as data-types and bounding values. The

## 7.2. MANAGING LARGE SYSTEMS

specified solution is Structure of Management Information (SMI) [CMRW93], which is a formal set of rules defining data encoding types. Common examples include:

- Unsigned32 – integer value between 0 and $(2^{32} - 1)$
- Counter64 – non-negative integer that increments positively, wraps to 0 at $2^{64}$
- Gauge32 – a bounded non-negative integer
- TimeTicks – time between 2 epochs in centi-seconds, $0 - (2^{32} - 1)$, also wraps

SMI also permits user-defined objects and sub-types to be created e.g.:

- Structures, defining value types and sub-types
- IpAddress – a 32 bit Internet address = a length 4 OctetString – and is an example pre-defined sub-type

SMI has it roots based in ASN.1 [ITU08], which is an ITU-T defined notation that allows data structures and values to be efficiently and unambiguously represented and distributed in a text format across multiple platforms. The most common and prevalent version in use today is SMIv2 [MPS+99], which is used together with MIB-II and SNMPv2c.

### 7.2.7 Previous Large System Research with SNMP

**Cluster Machine**

In [AMG04], Alves et al. discuss their implementation of a private enterprise MIB to manage their own cluster environments. They take the idea of removing the SNMP agent from each of the end stations, and having a front end to the many processing elements being handled by a 'cluster controller', however they limit their functions to a single management station which reduces both the capacity and the resilience of their solution. Their implementation was a success, but it did not take into account hardware monitoring other than just a processor percentage, memory utilisation, etc, as they were focusing on allowing an operator to be able to assign jobs manually to cluster machines with lesser loads. The Alves' solution does not take into account massive parallel processing solutions that have become increasingly prevalent, nor does it take a broader overall view of hardware and software management. It appears that the solution has been built to solve a particular sized problem rather than being created to cover a broad range of applications, and larger scale systems.

**Grid**

Subramanyan et al. [SyMAF00] describe an approach for much larger, albeit heterogeneous, systems – such as those undertaking grid type computing applications. In their SIMONE solution they describe an Intermediate Level Manager (ILM) function, which serves some of the consolidation needs in a very large scale system. They indicate heavy CPU usage as the main driver for this consolidation function, but do not approach the issue of reducing this CPU load at source, only to distribute it. They include a useful study of processor and communications overheads as part of their paper. Their work quite clearly demonstrates the benefits of distributing the management load amongst multiple agents unlike Alves et al. [AMG04], and they list AgentX [DWEF00] as one such method of reducing overheads of processing SNMP and MIB.

## 7.3 Summary

This chapter has explored the topic of managing large scale computing systems to maintain and control them. There are a number of aspects covered by the term 'system management' including faults, configuration, accounting data, performance and security. For any network attached equipment, the network itself can be used to provide the path between the managed and managing systems, and this enables remote and centralised system management to take place. For network attached systems there has been standardisation around the *protocols* used, but there are a wide diversity of management tools used by operators. System management tools typically provide a dashboard representation when supporting large systems, using positional and colour information to indicate the status of the items and faults being monitored. A hierarchy of management is also often supported, which removes complexity from the management screens, only displaying relevant detail. This concentrates the attention on the components with issues, so that highlighted issues are resolved as quickly as possible after the initial alert.

In the case of SNMP the data to be serviced is stored in the tree-like MIB database, whose structure is distributed amongst the managed agents and network management system(s). The MIB may be polled either to retrieve information, or to set parameters which remotely influence the behaviour of that end device. The structure of the MIB data stored on end devices is essentially arbitrary but hierarchical, and is related to the specific management requirements of that device. SNMP is deployed almost universally in remote system management where the systems connect to a network, and this

## 7.3. SUMMARY

has led to a wide-ranging set of tools to be developed to manage such systems. These tools provide multiple arbitrary perspectives for large systems, and may be enabled by centralised or distributed polling systems.

In the next chapter the system management of the SpiNNaker hardware architecture is tackled, and a framework developed to permit management data to be transferred to and from the system efficiently using standardised tools (which would normally be too heavyweight for SpiNNaker's constrained operating environment).

# Chapter 8

# SpiNNaker Management Framework

Monitoring is now almost ubiquitous for computing platforms, as the increased visibility permits remedial actions to be automated, and maintenance to be minimised. The visibility of both machine and software health (in real-time) is an important aspect of their ongoing operational and performance management. Providing real-time visibility of system resource loading on a machine consisting of many tens of thousands of components is not a trivial task and the ability to be able to detect and map around faults, together with the gathering of load information, permits dynamic allocation of work, re-routing, or further diagnostics to be initiated.

SpiNNaker is a novel, high-performance architecture, formed from large numbers of highly-interconnected, energy-efficient processing elements more typically found in embedded systems. This chapter presents the implementation and results of a management strategy for the SpiNNaker hardware environment (fig. 8.1), utilising a universal translation layer: SpiNNmate. SpiNNmate is a function located between a SpiNNaker machine and the communication protocols of external applications. On the SpiNNaker side the management strategy aims to minimise computational impact, while on the other side providing translation facilities to external protocols and tools.

## 8.1 SpiNNaker – a Memory-Mapped Architecture

SpiNNaker is a scalable, massively-parallel, computing architecture constructed from interconnected SpiNNaker chips, each containing 18 ARM9 processors clocked in the low hundreds of MHz. Each of the cores within the SpiNNaker chip has access to both individual and chip-level memories, and to the shared hardware of the system in a memory-mapped architecture (fig. 8.2). Each core locally has its own 32 kB and

## 8.2. A PROTOCOL TRANSLATOR – SPINNMATE

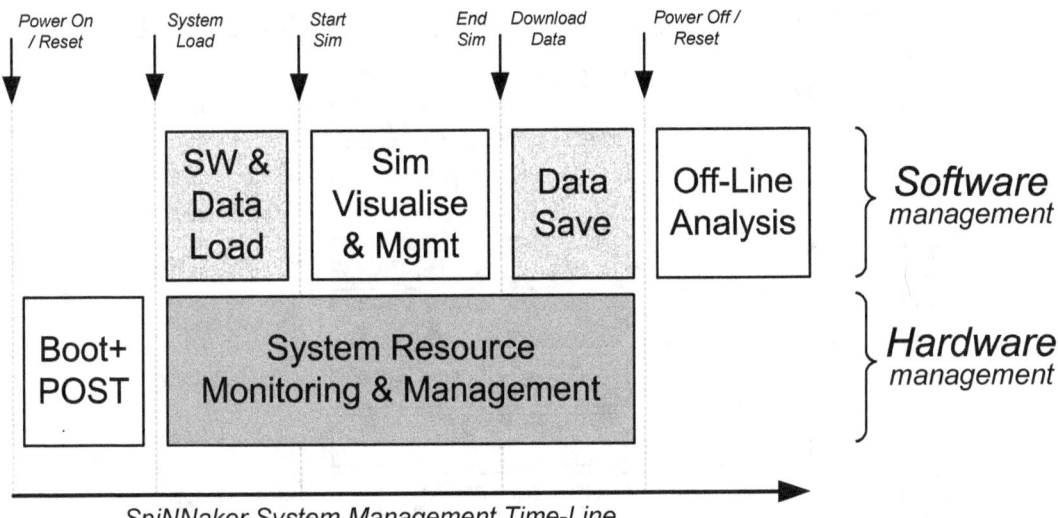

Figure 8.1: The system management time-line of the SpiNNaker platform – management of system hardware, peripherals and memory is carried out by the framework provided around SpiNNmate.

64 kB of instruction and data memory, and all cores have access to a shared system-level RAM of 32 kB, and to an off-die (but in-package) shared SDRAM of 128 MB. Both shared memories have multi-cycle latencies, but each core has DMA hardware support to transfer blocks of shared memory back and forth efficiently. In addition to the memories, the majority of the peripheral blocks at both individual processor and system levels are accessed via memory-mapped locations (fig. 8.2).

There are multiple audiences for real-time visibility of machine monitoring. Those who monitor hardware operations need information about system health, and end-users: their application's performance. Due to SpiNNaker's resource-constrained execution environment, implementing a full universal management suite on all cores is not practical, and would divert resources from the applications it is designed to support. To minimise the impact on SpiNNaker a low-overhead, but scalable, management framework has been created based on low-impact *GET* and *SET* operations which can access hardware and software data via the SpiNNaker memory map.

## 8.2 A Protocol Translator – SpiNNmate

It is unlikely that users of the SpiNNaker system will attempt to monitor it by accessing specific memory locations, they will use higher-level software tools. These tools may be bespoke, but in other cases standardised software and protocols may be employed.

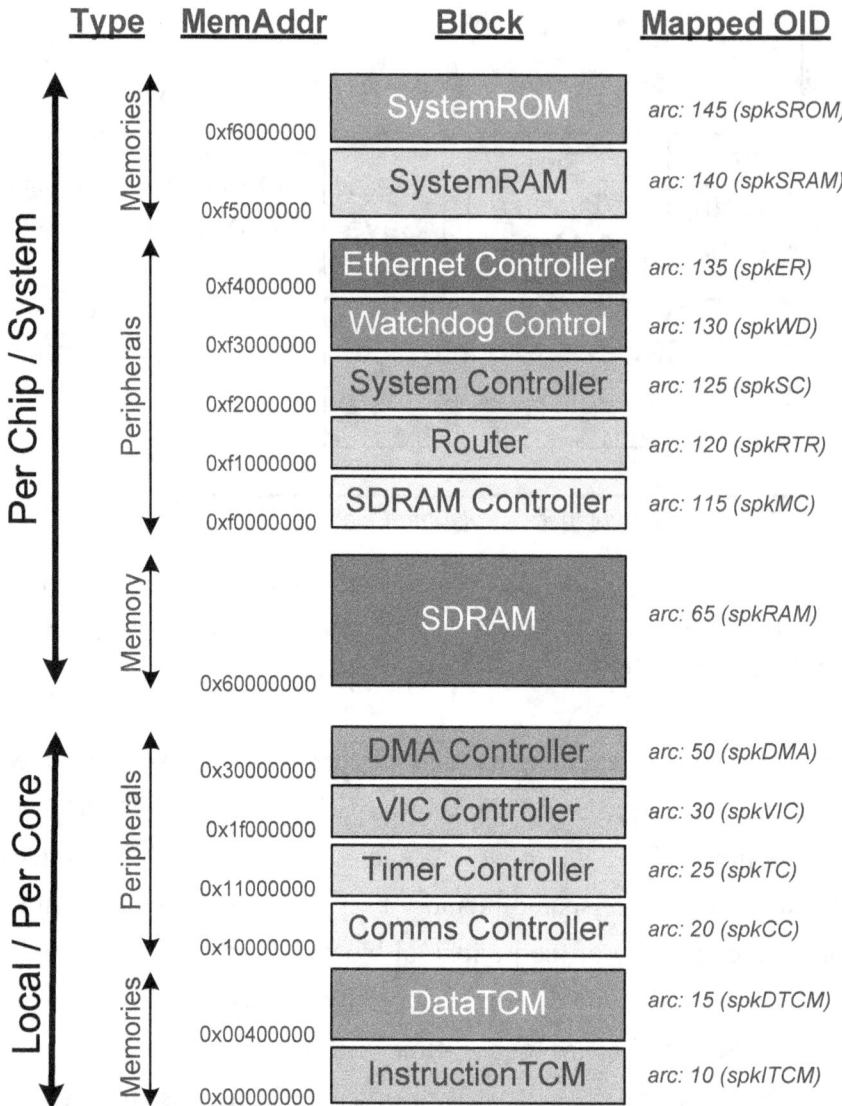

Figure 8.2: Memory map of a SpiNNaker MPSoC. This is mapped to a Management Information Base (MIB) model of the system with Object IDentifiers (OID) arcs indicated alongside each resource of the SpiNNaker MPSoC (see section 8.2.3).

SpiNNmate (fig. 8.3) is an intermediate management software layer providing a simple, extensible, framework for interfacing with the SpiNNaker platform. This approach leaves the majority of the complexity and overhead outside SpiNNaker with a set of lightweight, unified, primitives providing the I / O and control required of the SpiNNaker machine. SpiNNmate's primary target is to provide a common intermediate point of access for external protocols and applications that wish to communicate with SpiNNaker, and dispense with the need to implement bespoke code on SpiNNaker for each new application or provide on-board translation.

## 8.2. A PROTOCOL TRANSLATOR – SPINNMATE

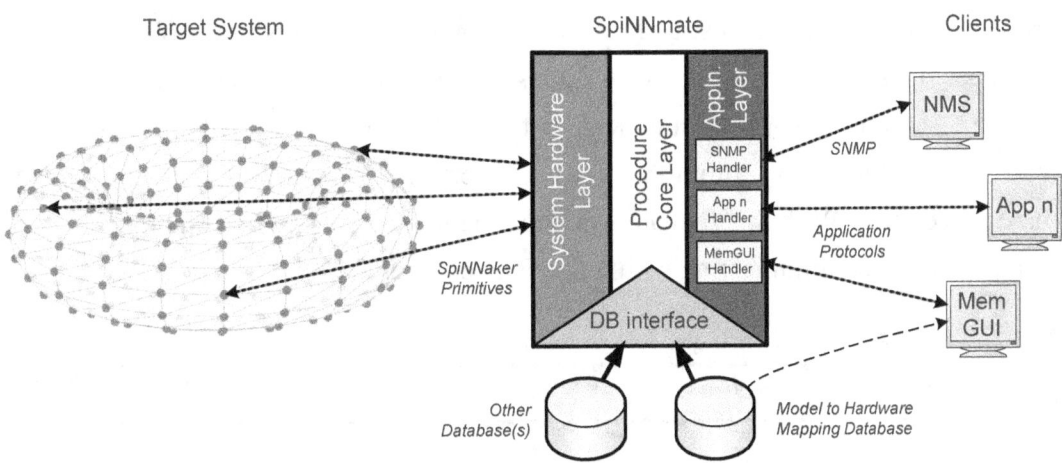

Figure 8.3: The SpiNNmate translation service. Multiple diverse applications / protocols (right) may access SpiNNaker via the translation layers of SpiNNmate.

### 8.2.1 Primitive Operations on SpiNNaker

SpiNNaker is a memory-mapped architecture; the vast majority of on-chip resources can be accessed with a simple read or write of the relevant memory location / memory-mapped register (fig. 8.2). These two simple 'GET' and 'SET' operations cover the majority of transactional requirements to control and manage a SpiNNaker chip. Accessing the system using these simple primitive operations ensures that the end processor does not become overwhelmed or have its limited instruction memory filled with operations that are rarely used. If operational complexity is required, this can be performed externally, formed by a series of GET or SET operations on the relevant memory locations. For this reason, the SpiNNmate framework has been developed providing GET, SET and a third primitive *RUN* which provide full access for control and system management purposes to all areas of the SpiNNaker machine.

### 8.2.2 SpiNNmate Structure

SpiNNmate is a protocol translator typically implemented on a external Host workstation which 'mates' external protocols with the resource-constrained SpiNNaker machine. It is formed of several layers providing a full translation service between primitives and different protocols and abstractions. At its first layer (leftmost 'System Hardware Layer' in fig. 8.3), SpiNNmate communicates with the SpiNNaker machine via Ethernet, dealing with system primitives encapsulated in packets (fig. 8.4). Next, SpiNNmate provides a core layer which operates with commands that are macros of system primitives, and provides abstractions via the database to SpiNNaker objects (rather than packets, primitives and memory addresses). The next layer, the application layer, is where user applications communicate with SpiNNmate. SpiNNmate provides translation blocks (handlers) within this layer, converting different protocols into SpiNNmate procedures. With SpiNNmate's modular nature, a new external client application can quickly be communicating with a SpiNNaker machine.

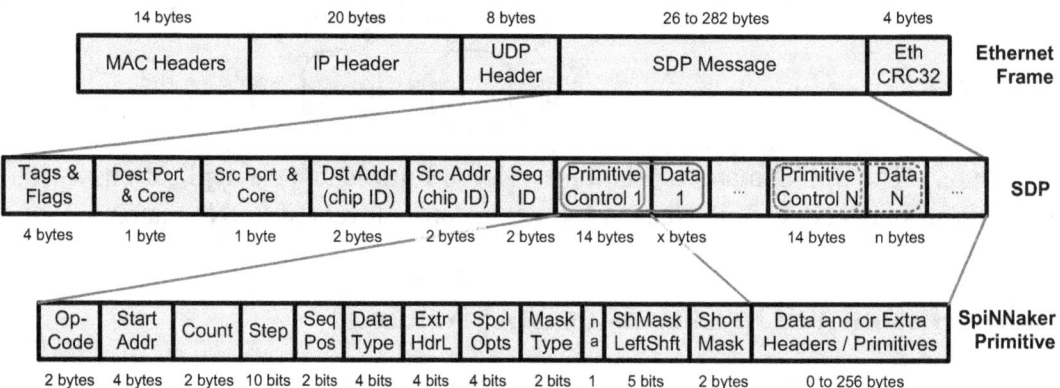

Figure 8.4: Communication between SpiNNaker and SpiNNmate is via SpiNNaker Datagram Protocol (SDP) packets. Within the SDP packets, SpiNNmate primitive(s) may be sent to / received from any SpiNNaker core in the machine.

**SpiNNmate System Hardware Layer**

SpiNNmate communicates with SpiNNaker chips over a network link using SpiNNaker Datagram Protocol (SDP). SDP is a transport protocol utilising standard Internet Protocol (IP), and provides a command space plus up to 256 bytes of payload data to be sent within its packet structure (fig. 3.7). The SDP header supports control and addressing (so that any core in a system may be addressed), and a port to identify target applications. The SDP data portion is prefixed by a command sub-header, which includes fields for sequence, command code and arguments. In practice this means that smaller SpiNNmate primitive messages can be conveyed without requiring use of the optional 256 bytes of optional data payload. SpiNNmate communicates with a small module of code on a SpiNNaker core which is listening for SpiNNmate primitive packets targeting it (fig. 8.4 details the packet format used). Within the command header three core operations (SpiNNaker primitives GET, SET and RUN) are encoded in the Op-Code field, with other fields including target memory address, repeat count, step-size, data type (words, shorts or bytes), and a mask that can be applied to the data.

## 8.2. A PROTOCOL TRANSLATOR – SPINNMATE

The RUN primitive includes additional headers optionally permitting scheduling and to provide regular reporting functions or alarms if a threshold is breached. Facilities to read or write large blocks of information from a single command may be encoded. The SpiNNmate hardware primitives layer also makes transparent any packet splitting, should the user request to GET / SET quantities of memory that exceed the SDP payload size.

### SpiNNmate Core Layer

This layer contains wrappers around the hardware primitives, procedures easing some of the system specific variable handling, and macro functions providing repetition of hardware primitives – e.g. supporting dumping and loading memory to and from files. An important part of this layer is that it can communicate with an SQL database containing the abstraction between 'objects' and the detail of SpiNNaker memory addresses (as well as bit masks, lengths, access rights and register descriptions). Therefore if a user requests an object (via one of SpiNNmate's protocol handlers), they are abstracted from the raw address and bit-wise detail. The database sits behind a DB interface which is available across all SpiNNmate layers.

### SpiNNmate Application Layer

The application layer handles the conversion of various protocols into SpiNNmate core commands. There are currently two translation blocks supporting management applications, which are expanded in section 8.2.3 below. In operation, a protocol block listens for its packets, interprets them into data structures, and makes a request to the next layer of SpiNNmate to retrieve one or more objects, or to write certain values to them.

### Proactive Alerting

By running a scheduled service routine on the Monitor Processor, chips may proactively send out alerts to the management system when specific registers / counters breach user-defined thresholds. This mechanism avoids the necessity to continually poll this information from outside the system, and thus saves computing and bandwidth resources.

### 8.2.3 Protocol Modules

Two different protocol handlers have been implemented for management applications, with very different traffic profiles to meet the hardware system management and file operations requirements of figure 8.1. The first application is to interface SpiNNaker with SNMP, which is a de-facto standard for monitoring and management of network connected devices. By implementing an SNMP protocol module in SpiNNmate, the SpiNNaker machine is opened to dozens of standard system management tools which can be customised for monitoring, alerting, performance and fault management. SNMP is predominantly a transactional protocol, and is used as such in the experiments. The second implementation is a GUI which interfaces with SpiNNmate to manipulate the memory / hardware on a SpiNNaker machine. Combinations of primitive commands are used to GET and SET large blocks of data on the shared memory of each chip pre- and post-simulation. This communications methodology covers the data-loading and results-saving requirements of simulations executing on the SpiNNaker system.

**Implementing SNMP and Constructing SpiNNaker's MIB**

As described in chapter 7, Simple Network Management Protocol (SNMP) [CFSD89] is designed to be a simple, low overhead protocol permitting Network Management Systems (NMS) to set and retrieve information from multiple network attached elements. The management information such as settings, statistical counters and gauges is stored on the managed agents in the Management Information Base (MIB) structure. The SpiNNaker MIB is formed of specific SpiNNaker objects that the agent maintains, and can be accessed by the NMS.

As each SpiNNaker chip (and core) does not have its own IP address and stack, it cannot run a typical full SNMP agent. Only a subset of SpiNNaker chips are provisioned with Ethernet connections which are assigned IP addresses. Potentially therefore, an Ethernet-attached chip could act as master agent, with all other downstream processors acting as slaves in an AgentX implementation. Two problems arise with this option: AgentX is an SNMP specific extension whereas the SpiNNmate framework is agnostic to the input protocol, and the overheads of running AgentX clients are higher than the simple primitive operations outlined. Therefore, the SNMP agent is implemented by SpiNNmate itself, so an NMS directs its SNMP messages at a SpiNNmate protocol module.

## 8.2. A PROTOCOL TRANSLATOR – SPINNMATE

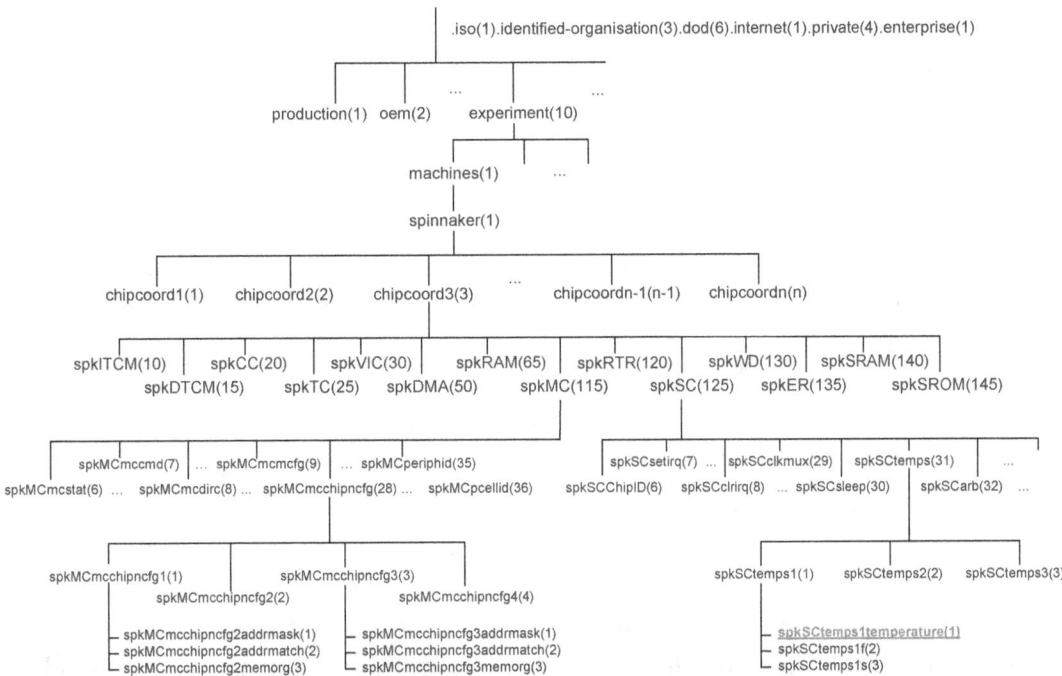

Figure 8.5: Extract from the SpiNNaker MIB tree structure. Most of the subtrees are pruned in this diagram so that examples from the Memory (MC) and System Controllers (SC) may be highlighted.

MIB objects are required to be globally unique, and are referred to by their OID (Object IDentifier) which is either in dotted decimal notation, or an associated textual representation. The connectivity of SpiNNaker provides a unique co-ordinate address for each chip, and together with SpiNNaker's memory map (fig. 8.2), this forms the template for the construction of the SpiNNaker MIB structure.

The MIB source of a SpiNNaker machine begins with its chip co-ordinate, each chip having beneath it an identical MIB branch structure. This identical structure is derived from the memory map of SpiNNaker, with numeric and text OID arcs (subtrees) defined for each distinct block (see the right-hand side of fig. 8.2). Each block may then branch to registers, then each instance of that register, and finally any subregisters beneath that.

To illustrate this, a subset from the SpiNNaker MIB is reproduced in figure 8.5. To navigate to the temperature sensor 1 leaf object (highlighted in the lower right of the diagram), either the numeric or text OID may be used. The temperature value has a full numeric OID string: *.1.3.6.1.4.1.10.1.1.3.125.31.1.1* or may also be represented in text as: '*.iso. identified-organisation. dod.internet.private.enterprise.experiment.machines .spinnaker.chipcoord3.spkSC.spkSCtemps.spkSCtemps1.spkSCtemps1temperature*'.

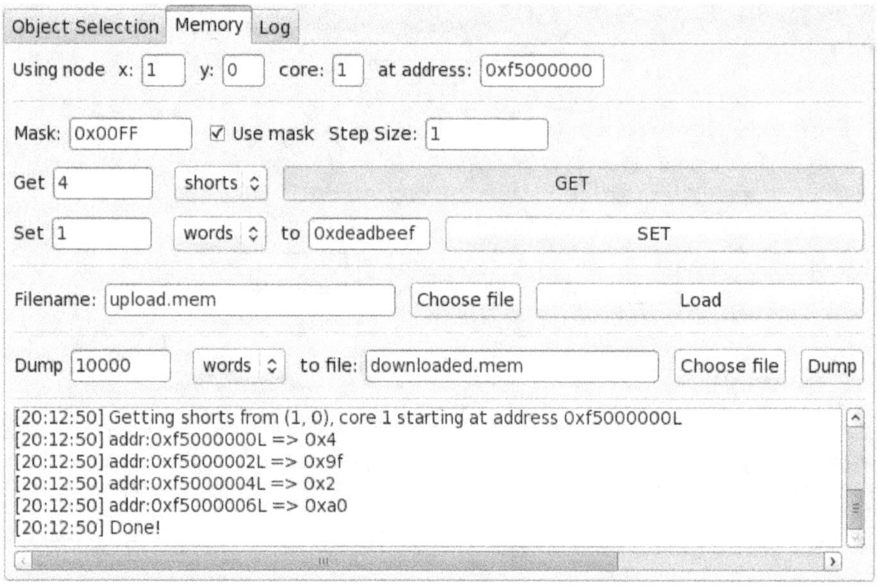

Figure 8.6: An example of MemGUI where the user may select to GET or SET address(es) in memory of a specific chip / core, with options to repeat, mask and to load / save files on the target SpiNNaker machine.

The information required to create the SpiNNaker MIB was created in a database and includes all data required to distil into the standard MIB format, including OID numbers, names, permissions, a free-form text description and its syntax. This same source database also includes (for use in the SpiNNmate hardware layer and the SNMP agent), the mapping between OID and SpiNNaker memory address, together with mask and bit positioning information (typically used by sub-registers). The MIB generated from this source database may be imported into SNMP management tools, and used to query SpiNNaker machines, chips and cores by object name as required.

**MIB Enterprise Number**   The University of Manchester does not own an enterprise number for use as a tree root within the proposed MIB. therefore the assignment of a new enterprise number may be appropriate, especially if commercial exploitation of the SpiNNaker Intellectual Property is sought. It would therefore be prudent in future to apply for a unique enterprise number for this project to create a unique enterprise sub-tree in the longer term.

**MemGUI Implementation**

MemGUI (an example of which can be seen in fig. 8.6), is an easy-to-use graphical method of gaining access to read and write a specific SpiNNaker chip's memory.

Underlying MemGUI's operation are a set of commands which are used to access operations from the SpiNNaker machine via SpiNNmate. To provision the MemGUI load and save operations, its commands are broken down into constituent primitives by the core and hardware layers within SpiNNmate. By providing macros of low-level operations, other more complex transactions may be created.

As the object to hardware mapping database is provided externally to SpiNNmate in a standard format, it may be accessed by more than one client. In this case the database created primarily for the SNMP MIB is used to provide MemGUI with a user-friendly method of accessing parts of the system by object name or description, rather than having to traverse the memory map by address (leftmost tab in fig. 8.6). The extra client linkage of MemGUI to the database is illustrated in figure 8.3.

## 8.3 Memory and Communications

Fundamental to the implementation of SpiNNmate and its primitive GET and SET operations are the performance of the memory operations that it carries out, and the communications of the requests / results to the SpiNNmate platform. A number of experiments have been carried out on both categories of infrastructure to maximise the potential of each of these paths when conveying management information.

### 8.3.1 Memory Operations – When to use DMA

SpiNNaker has a multi-level model of memory – the view from each processor across a system is incoherent, with no memory visibility beyond the local chip package. Each core has its own relatively small (in the tens of kilobytes) instruction (ITCM) and data (DTCM) memories, private to each processor block (fig. 8.2). Further and larger store is shared by all processors at the chip / system level, and accessed over the System NoC, which each processor may elect to access using its DMA controller.

As the GET and SET primitive operations are fundamentally memory operations, memory performance is of key concern to the SpiNNmate platform. This is particularly the case where large sections of data are moved about, such as loading and saving data on the SpiNNaker system pre- and post-simulation. Therefore experiments were performed between blocks of shared system and local processor memories to understand their characteristics, and to determine the point at which it becomes more efficient to access shared system memory by DMA rather than use item-by-item operations.

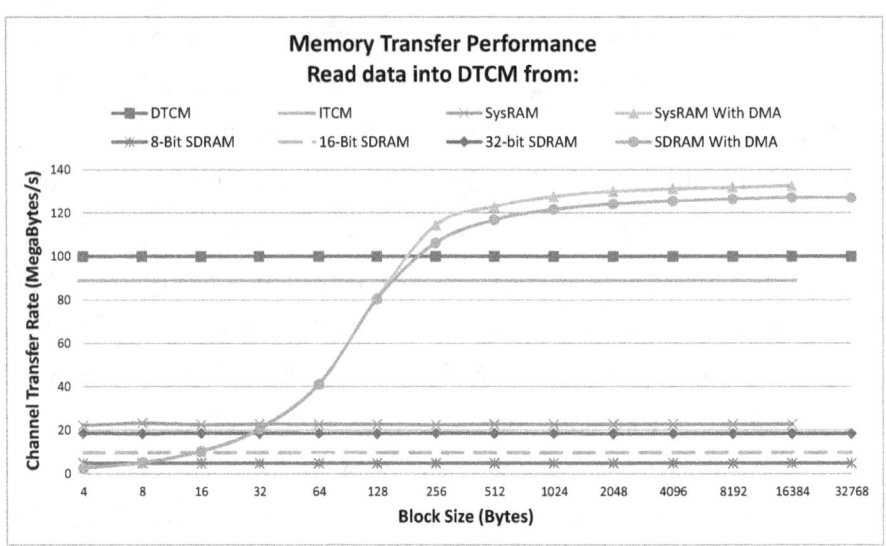

Figure 8.7: Comparison of memory read performance using on different types of memory in the system. Selecting DMA on all but the smallest block sizes is faster.

A number of memory performance tests were carried out in the standard operating environment of the system; this is a thin API which abstracts from the direct hardware, and is the environment in which SpiNNmate memory transfers are performed. Each test consists of a repeated gigabyte transfer of data, and the elapsed time is recorded by reference to the 2nd on-chip timer at the start and end of every transfer (microsecond resolution). Results from the testing are detailed in figure 8.7, where different memory block sizes are transferred to and from DTCM using all four types of memory (including DMA where available). Results for read and write figures were approximately comparable, therefore write results are omitted from the figure for clarity. A 32-bit word size was used for most transfers as it is consistent with the ARM 32-bit architecture. However, a user may wish to access memory at other item sizes (chars, shorts), therefore experiments were also performed with 8- and 16-bit operations for comparison.

The results of the memory tests indicate that a DMA memory operation should be initiated where the size of the transfer to / from shared memory is 32 bytes or larger. For shorts / bytes the crossover occurs sooner, as the number of instructions scales proportionally to the number of memory accesses. Using DMA for system memory has an additional benefit not recorded in these results: that the processor may undertake other tasks whilst awaiting the response from the DMA controller. Therefore in the implementation of the SpiNNmate responder, SpiNNaker has been coded to switch to DMA as indicated by this set of results.

## 8.3.2 Memory Performance Optimisation

A further set of experiments were performed on the tuning of the DMA controller parameters used for the memory transfers. The output of this work was the selection of parameters to be used for DMA transfers within the API environment. The bulk of the experiments were carried out between the DTCM memory of a core and shared SDRAM memory (the predominant main-store path) to optimise these transfers. The method to record results used in the earlier tests was used again here. The traffic passes between the (separately) clocked domains of the requesting core and shared memory, traversing the asynchronous System NoC (fig. 3.2), therefore any poorly performing regions as a result of interactions or 'beating' are of particular interest.

**DMA Parameters**

The two parameters which may be chosen to tune the DMA transfers are the width of the transfer (1 or 2 words (32 or 64-bits)), and the burst size (which may range from $2^0 - 2^4$). Figure 8.8 presents the results of write operations using the same experiment as earlier. The benchmark for results are those recorded in figure 8.7, where the maximum transfer rate achievable was ~125 MB / s. It can be determined, almost across the board, that a burst size of 16 together with a double-word width produces the best performance, particularly on larger blocks where an approximate tripling of performance is achieved. The outcome of these experiments is that these parameters have become the new defaults for the DMA transfers in the API.

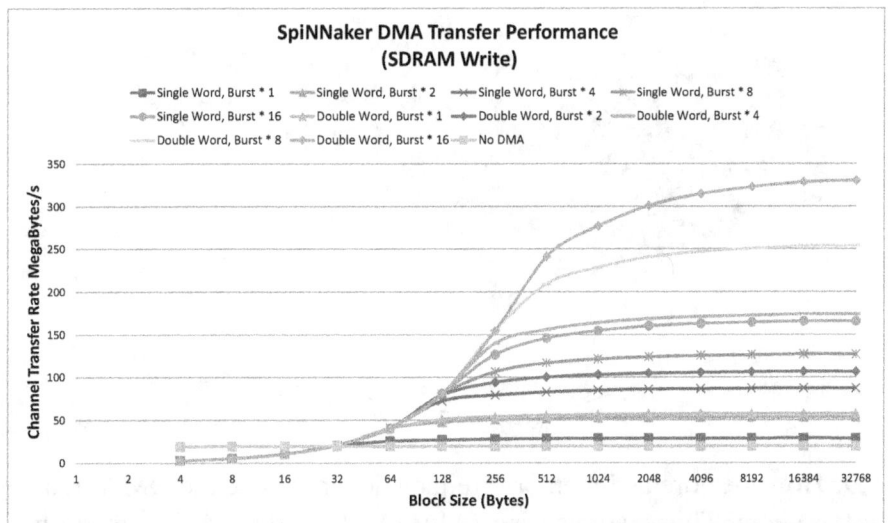

Figure 8.8: Comparison of memory performance using various DMA parameters, larger burst sizes of double words typically provide the best performance.

### Clock Speeds

This next experiment explores the relationship between the clocking domains on the SpiNNaker chip, and how they influence memory transfer performance. There are five clock rates which may be influenced by the user. The processors are split into A and B domains (9 in each, so clocks may be differentiated if desired). Two independent clocks are derived for the router (not relevant in this test as there are no packets to be routed), and the System Bus (which again is not relevant as the path to the SDRAM does not pass over this bus (fig.3.2)). The final clock is supplied to the memory controller to drive the SDRAM.

The experiment therefore uses the 2 degrees of freedom: core and memory clocks to create a performance surface (using the DMA burst parameters determined as best in the previous experiment), and to identify combinations that perform well, and eliminate any which are poor performers.

The results for read and write directions in this experiment are quite different. The write direction is the easiest to interpret (fig. 8.9), with its smooth surface contours. The limiting factor below a core clock rate of 180 MHz is the performance of the core itself, and above this knee only a very minimal increase is available, so the limitation is elsewhere. Raising the memory clock rate slowly, but steadily, provides better performance for the same clock core rate. The ceiling therefore appears to be located in the channel capacity to the fabric.

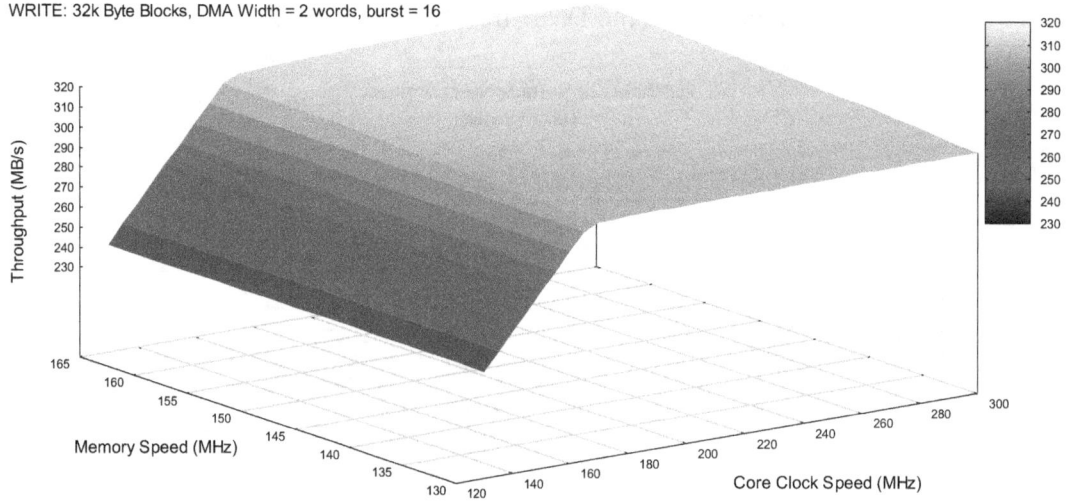

Figure 8.9: How the core and memory frequencies influence the DMA transfer rate to SDRAM for a core. This example uses 32 kB blocks and the write direction.

## 8.3. MEMORY AND COMMUNICATIONS

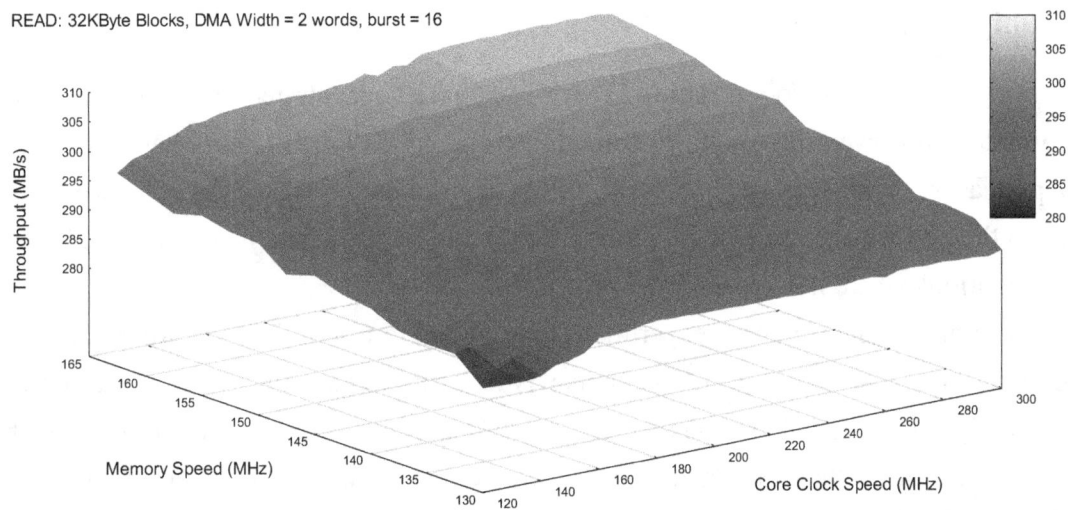

Figure 8.10: How the core and memory frequencies influence the DMA transfer rate to SDRAM for a core. This example uses 32 kB blocks and the read direction.

The read graph is quite different from that of the write direction, although it still exhibits trends. The resulting surface plot (fig. 8.10) demonstrates that at core clock speeds below 160 MHz the performance of memory reads drops off relatively quickly, with the memory clock rates not playing a part. There is a performance plateau that appears at a core clock of 240 MHz, and anything above this does not gain any significant extra memory performance. From the memory clock perspective, the faster the memory speed, the better the performance achieved, rising almost linearly.

The exact reason for the uneven read performance requires further investigation, but is likely to be a result of the internal scheduling performed by the memory controller in the read direction (attempting to optimise throughput), whereas in the write direction the operations are carried out strictly in order. To validate these results the experiment was repeated across all combinations of burst parameters and block sizes, with the read and write surfaces maintaining the shape discontinuity seen in the illustrations.

To give these results some context the SpiNNaker chip is expected to be clocked at 200 MHz core speed and a 165 MHz memory clock in production. By maximising the core clock from these notional rates a <1% throughput increase in the write direction is observed, and for the read direction an improvement of 1.3%. This experiment proves that inflating core speeds beyond their notional rates for purely memory performance purposes does not provide a sufficient pay-off versus the significant power increase required.

## Multiple Cores

Thus far, the memory optimisation experiments have concentrated on single-core performance, however, this is not expected to be operationally realistic – there are many Application Processors all expected to utilise the shared memory. In this final experiment the performance and fairness of the shared access fabric and memory is explored, in situations where it is contended.

The previous throughput experiment is extended to run on incrementing numbers of cores, and a synchronisation mechanism is set up between them. The arbitration tree of the System NoC (fig. 8.11) is used to determine a 'fair' order in which to add each additional core to the experiment. Firstly seven cores are added one-by-one in the read direction, followed by seven write cores, also added one-by-one. The physical core order is: 6, 4, 2, 8, 14, 12, 16, 1, 7, 5, 3, 9, 11, 13 which avoids the elected Monitor Processor on that chip and the 'top level' asymmetric shares.

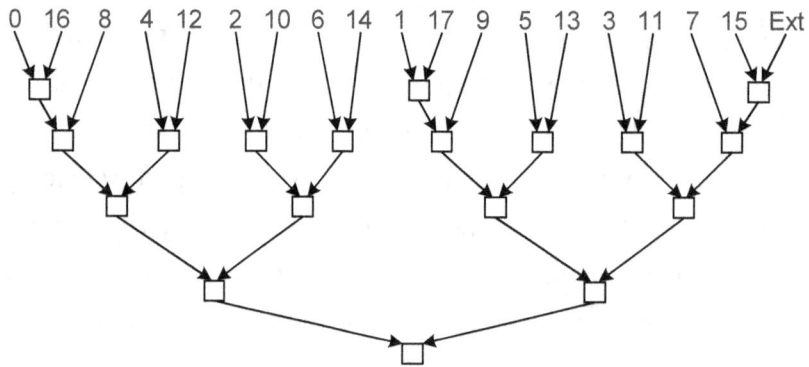

Figure 8.11: Accessing the System NoC the cores pass through an arbitration tree, each multiplexer a potential cause of contention / limitation.

The results of this experiment show that fair proportions of the bandwidth continue to be apportioned to each of the cores as expected. This is graphically displayed in figure 8.12, which shows that around ~600 MB / s is available to be read and up to 900 MB / s can be recorded when traffic is flowing bidirectionally over the fabric. Where the tree becomes unbalanced then each core sharing that branch of the tree still has a fair share of that bandwidth. For example where there are 12 cores in operation, there is less individual bandwidth available on cores 1, 5, and 9 in the write direction than 7 and 3, as there are 3 cores sharing one path in the multiplex tree versus 2. There is some disruption in overall data rates when write requests are simultaneously operating, although the bandwidth shares are always fair, further investigation is again required to profile all components in the system path including the memory controller.

## 8.3. MEMORY AND COMMUNICATIONS

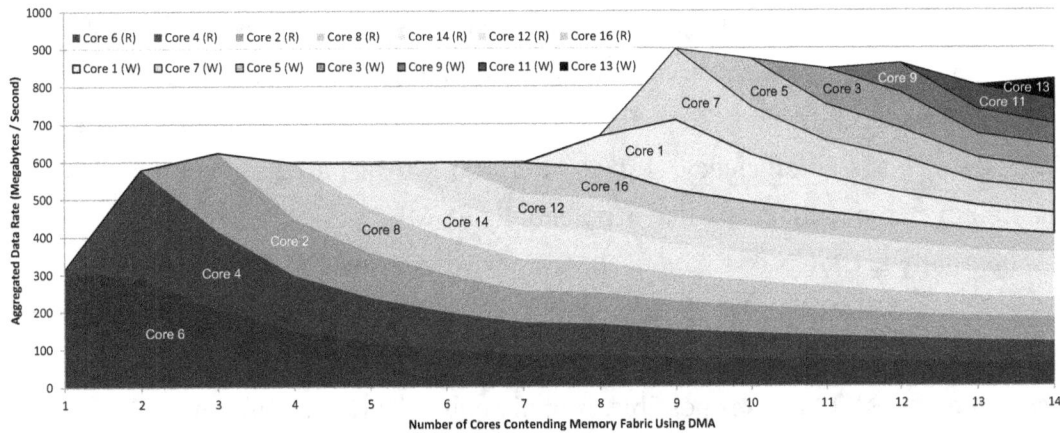

Figure 8.12: One by one cores are added transferring data with DMA, the first seven performing reads, then the following seven writes to maximise use of the asynchronous fabric of the System NoC. (Core clocks all set to 200 MHz and memory 165 MHz).

### 8.3.3 Communications Bandwidth

**Hardware**

By generating packets on local cores it is possible to achieve 5.3 Gb / s at entry to the router, shared amongst the cores by the input processor merge tree. A single (overclocked) core in a tight loop can individually achieve 5.0 Gb / s. Each off-chip link is capable of sustaining around 250 Mb / s, a figure limited by the on- and off-chip delays.

The Ethernet hardware is capable of supporting 83.2 Mb / s incoming, and transmitting at 92.8 Mb / s. These figures are based on raw packet rates (including headers and trailers) and do not include any processing overheads or checksumming that may be required to be productive with the network data.

**Software (SDP)**

In the current small SpiNNaker machines, SDP is used to load application data and executable code. A measured test, transmitting data from a Host machine to an Ethernet attached SpiNNaker chip and then to an Application Processor on that node via shared memory, achieved speeds of 4.7 Mb / s. With the target Application Processor on a different chip the scenario now appears as per figure 3.8. Here the payload transmission speed is 3.9 Mb / s, due to the fragmentation, bridging and acknowledgements of the internal SDP using P2P packets.

The SDP datagram payload sizes used in the tests were 256 bytes. The total Ethernet frame length of an encapsulated 256 byte SDP packet is 328 bytes: 18 bytes of Ethernet Headers / Trailers, 20 of IP, 8 of UDP and 26 of SDP giving an overhead of 72 bytes. This is an efficiency of 78%. For 100 Mb / s Fast Ethernet therefore, the maximum effective SDP data rate is potentially 78 Mb / s.

Internally to SpiNNaker, P2P packets are used to convey SDP data between chips. Each 72-bit SDP P2P packet carries a 24-bit payload. After including the headers, this gives an efficiency of 31%. Given the ~250 Mb / s chip-chip link rate this suggests a peak data rate ~80 Mb / s is possible, roughly equivalent to the Ethernet.

The conclusion of these experiments is that memory access times are almost insignificant when it comes to communications performance. Network communications are clearly the biggest performance bottleneck for SpiNNmate primitive operations. The communications performance and potential improvements are discussed in greater detail in chapter 9.

## 8.4 Management Framework Results

In this section a number of experiments are presented using the SpiNNaker system management framework. Results recorded from both the block transfer memory GUI application and the SNMP model are provided.

### 8.4.1 Memory Operations – the Performance of SpiNNmate

The capabilities of SDP as the underlying SpiNNaker data transfer protocol play a large rôle in the data rate that can be achieved between SpiNNaker and an external management device. The following results focus on the end-to-end performance of SpiNNmate, comprising both memory and network components.

In the experiment GET and SET (read and write) primitive operations are invoked via SpiNNmate (using the MemGUI application). Figure 8.13 details the performance of two memory classes – System (SDRAM) and Local (DTCM) hierarchically comparing: native memory performance on-chip, Ethernet channel potential, SDP potential performance, and SpiNNmate results (for both an Ethernet-attached chip, and one that is a hop away over a SpiNNaker chip-to-chip link). The results are presented logarithmically in the $y$ dimension as there is a wide variation between operational results.

## 8.4. MANAGEMENT FRAMEWORK RESULTS

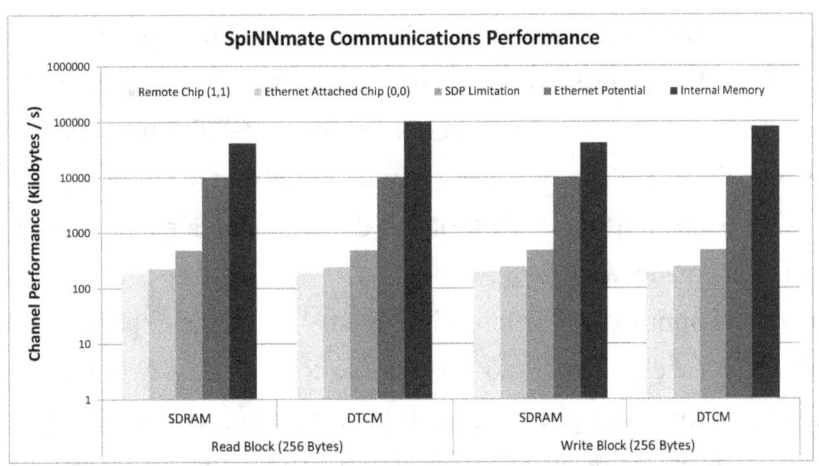

Figure 8.13: Performance comparison of memory and communications transfers in a SpiNNmate implementation.

When compared with the native memory and network bandwidths, the figures are disappointing. The rates achieved by SpiNNaker (2-300 kB / s, depending on target chip location) are many times slower than the native data rate of the 100 Mb Fast Ethernet connection (which is the slowest of the physical connection paths between all the devices). On further examination, however, they do bear scrutiny with the maximum SDP data rate that can be achieved over the Ethernet, which currently applies a limiting factor (measured at 588 kB / s to an Ethernet-attached chip, and 452 kB / s for a remote chip). SpiNNmate achieves around 50% of this available SDP bandwidth, but offers only a single packet in-flight at any one-time, so the serialised latency of a packet (and its processing) determines the maximum rate of transmission. It should be possible to almost double the rate and reach near the SDP notional maximum by having 2 packets in-flight at any one time.

**SpiNNmate's Added Features**

The SpiNNmate protocol includes the ability to mask, jump and repeat data for both GET and SET primitive operations. The impact of applying these features to attainable data rates has been examined, with the results provided in table 8.1. The *Repeated Data* feature is useful to load a memory block specifically with a value (e.g. resetting SDRAM). Variants include repeating with increments or decrements to the data, and repeating a whole block. *Step Size* provides operations in intermediate steps (hopping over $n$ addresses each time), which could be ideal to retrieve / reset variables in a repeated structure. *Masking* for the GET primitive involves retrieving data then applying the mask before transmitting the results, whereas for SET the operation is subtly more complex. With SET, if a mask is applied, existing data in the target location is retained if the mask does not cover it. This process requires a GET, application of the mask, then a SET with the combined (ANDed) result. The advantage of using SNMP objects in this situation is clear, as the SNMP object database contains the sub-register mask mappings (a ready made mask).

Various repeat options are explored in table 8.1, with the experiment target to reset all SDRAM on a chip. With no options the Repeated 64 Word Block is clearly the fastest method for this application, with fewest packets (8) and lowest overall data transferred to perform the reset. When applying options, little adverse impact is seen on overall data rates, although the impact is disguised somewhat by the slower network throughput. Internally the step-over performance drop is more dramatic as DMA cannot be used as operations are not on a contiguous block. Care has been taken in the implementation of SpiNNmate to ensure that, where DMA is beneficial and possible to use, it is deployed.

| Repeat Approach | Packets/ Data TX | No Options | Mask: 0x00078980 | Stepover: 2 (interleave) |
|---|---|---|---|---|
| Blocks of 64 Words | 524288) (172MB) | 185.696s (0.7 MB/s) | 190.848s (0.7 MB/s) | 206.150s (0.6 MB/s) |
| Repeat Single Word | 512 (37kB) | 13.319s (10.1 MB/s) | 13.320s (10.1 MB/s) | 13.350s (10.0 MB/s) |
| Repeat 64 Word Blk | 8 (2.6kB) | 1.177s (114.0 MB/s) | 1.182s (113.5 MB/s) | n/a n/a |

Table 8.1: Setting SDRAM (128 MB) via SpiNNmate and SpiNNaker primitives. Differing operational approaches and 'features' are compared.

### 8.4.2 A Real-Time SNMP Temperature Plotter

The SNMP translation functions of SpiNNmate are used for a practical application of a chip temperature plotter. The SpiNNaker chip temperature sensors are a good test-case as they are highly dynamic (although, as yet, not calibrated). Whilst thermal load is applied to a chip (by having the Application Processors execute looped instructions), SNMP was used to poll the temperature sensors of all chips repeatedly (see the SpiNNaker MIB extract in fig. 8.5). The temperature is a secondary effect of the load placed on the processors in the system, very much like fMRI brain scans use a

## 8.4. MANAGEMENT FRAMEWORK RESULTS

secondary effect to detect areas of activation by examining blood oxygenation levels.

Using the values retrieved by SNMP, a scrolling real-time graph for all four chips based on the temperature gauge outputs can be computed and plotted (fig. 8.14). The sensors in the SpiNNaker chips clearly show at time 275 s the temperatures drop (fingers were placed onto the chip packages for ~20s, helping conduct heat away), and this behaviour reverted at around time 295 s. A similar thermal impact can be observed by moving air over the chip packages. (SpiNNaker chips emit only around 1 W when all cores are under load, and are not actively cooled on the test boards (fig. 4.13)).

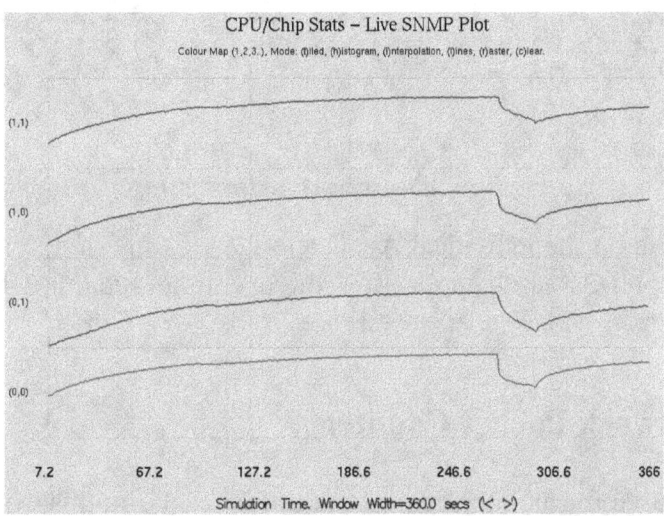

Figure 8.14: SNMP retrieved plot of 4 SpiNNaker chip temperatures when load is applied to all processors. At simulation time 275 s cooling is supplied to all 4-chip packages, and removed shortly afterwards. This plot uses the real-time visualiser developed in chapter 6.

### 8.4.3 Processor Utilisation Monitor

The SpiNNaker MIB and SNMP may also be used to poll all processors' activity on a chip centrally at any instant. A synthetic application was written which creates ramped loads on the Application Processors, with the load pattern shifting each 30 seconds. The status of each core was polled using SNMP with a mean rate of 3 Hz; jitter was added to reduce any synchronisation / beating within the data as the sample period is low compared to the 1 kHz load cycle. As each poll result is binary (0=active, 1=asleep), the results plot (fig. 8.15) has a rolling average calculated over a window of the previous 100 samples. From the figure the application load pattern can be discerned within the output data plotted for a number of SpiNNaker cores.

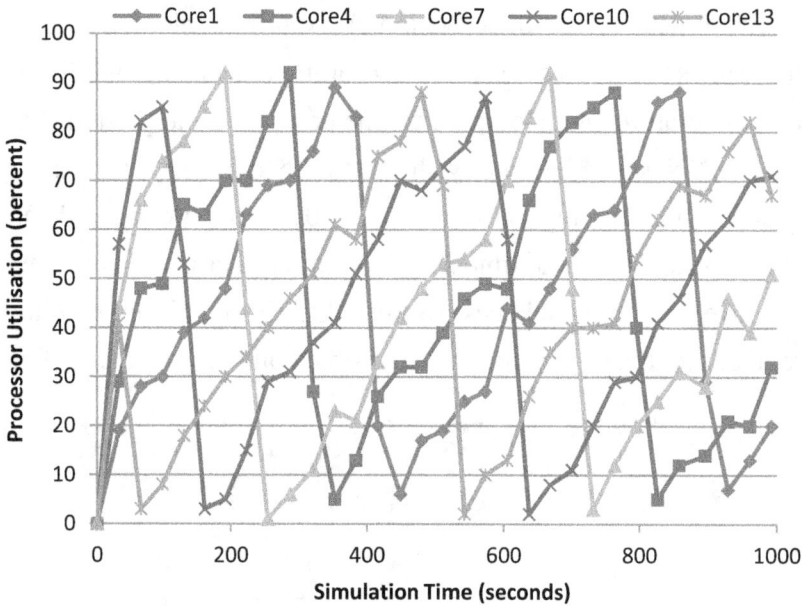

Figure 8.15: Load on the individual Application Processors of a SpiNNaker chip. A varying load is applied to each application core over time, and this data is retrieved via SNMP.

### 8.4.4 A Network Packet Counter

SNMP is often used for monitoring network traffic levels. In this experiment SpiNNmate SET commands are used to enable router counters for multicast packets which originate in each of the 4 chips in a large synfire chain [Abe82] (a spiking neural network simulation). Within this network, four populations of neurons are connected to one another and a cascade of events moves from one population to the next in a chain reaction. The trigger for the first population (a bias current) is enabled for 30 seconds and then disabled, and the results of this network show that the population firing pattern is offset in time due to its position within the chain.

SNMP is used to poll each population / chip's packet counters, and the spike firing-rate in Hz is plotted in figure 8.16. Here the temporal offset between the populations on each chip can be seen – and the ordering of the chain by chip may be deduced.

### 8.4.5 Long Term Monitoring

An extended test was carried out on a quiescent SpiNNaker chip over a number of days, using the SNMP protocol to perform trend and performance monitoring using an extended sample period (minutes). The test uses the most dynamic temperature sensor of

## 8.4. MANAGEMENT FRAMEWORK RESULTS

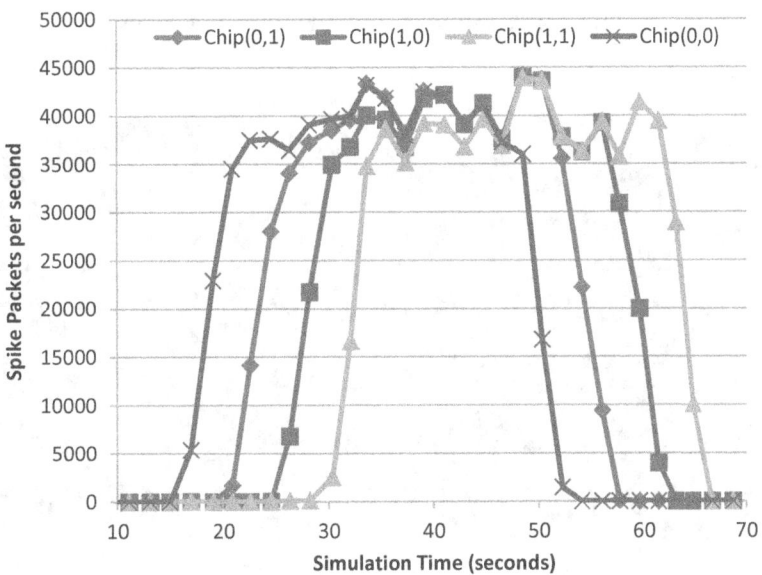

Figure 8.16: Packet / spike-rates per second generated by a SpiNNaker multi-chip synfire chain neural simulation and retrieved via SNMP.

the three on the SpiNNaker chip as a data source, which reacts to give a lower reading when the ambient temperature increases (in the experiment the cores are quiescent). The results were inadvertent, as the test was intended to be a week long soak / stability test of SpiNNmate, and the SNMP protocol module. The results (fig. 8.17) clearly capture environmental data about the room the board is located within – showing the lab heating system is turned on at 7 am daily, and that the heating is not operated at the same level over the weekend (4 – 6th February 2012). The temperature in the lab prior to the weekend was remarked upon as being uncomfortable, and to compensate the radiators were altered downwards during the afternoon of Friday the 3rd February. It is possible to see that the ambient temperatures were not as high in the weekdays following the adjustment.

The figure 8.17 graph was captured from the open-source SNMP management tool Cacti, which queried the temperature gauge every 5 minutes during the monitoring period. The results show that the SpiNNmate SNMP protocol module and responding code on the SpiNNaker chip are reliable over extended periods, and secondly that, by calibrating the temperature sensors, it will be possible to read the operational temperatures of the system – and detect the secondary effects of diversely active neural activity in a system. If neural models are properly mapped in simulation it should be possible to create a convincing real-time fMRI imaging equivalent on a SpiNNaker machine, also based on this secondary effect (temperature).

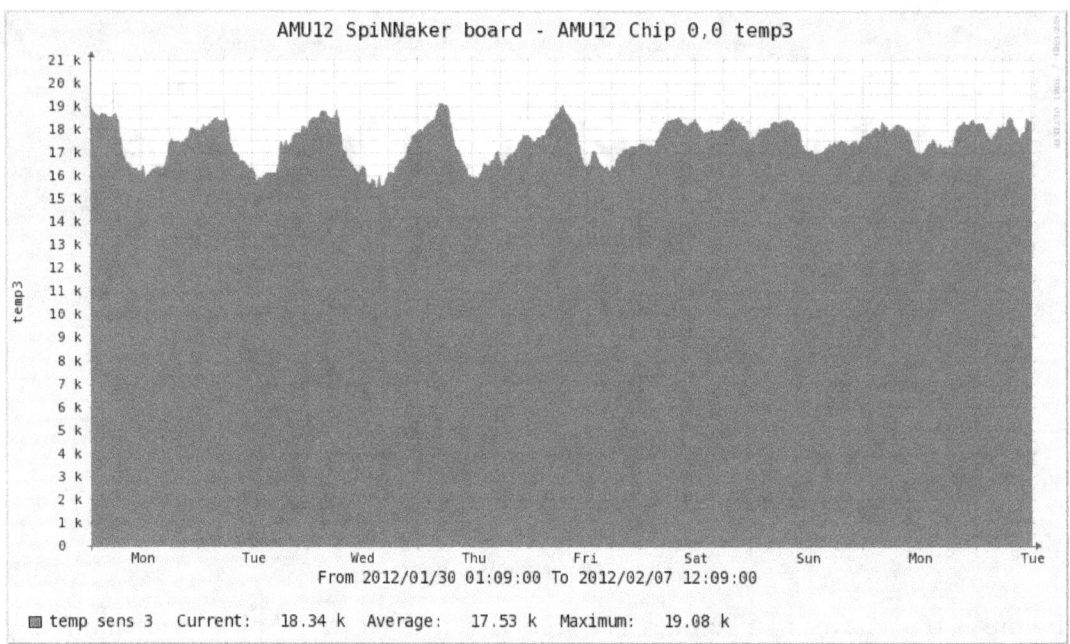

Figure 8.17: The Cacti system's plot of a SpiNNaker chip's temperature sensor data over a weeklong period. A lower reading indicates higher ambient temperatures.

### 8.4.6 Alerting using Nagios

SNMP is used (via SpiNNmate) to periodically poll all three of the temperature sensors of the chip being tested, from the widely-used Nagios system management tool [NE12]. Nagios has been set up so that if the temperature wavers upward or downward outside certain arbitrary ranges, alerts will be raised to warn operators of the breached temperature thresholds, and when exceeding a second set of more extreme thresholds, creates a critical alert for high and low temperatures.

In figure 8.18 a number of status alerts for the SpiNNaker board can be seen, as the temperatures rise and fall repeatedly inside and outside the monitored bands. The temperatures are being influenced by the application and removal of load to Application Processors on the test board being monitored.

This trivial threshold alerting example can be extended beyond temperature sensors on the chip, to different system objects for example traffic levels, core utilisations and error counts, such as those seen in previous experiments, or indeed any object which can be read from the SpiNNaker MIB tree.

## 8.5. MONITORING A SPINNAKER MACHINE IN PRODUCTION

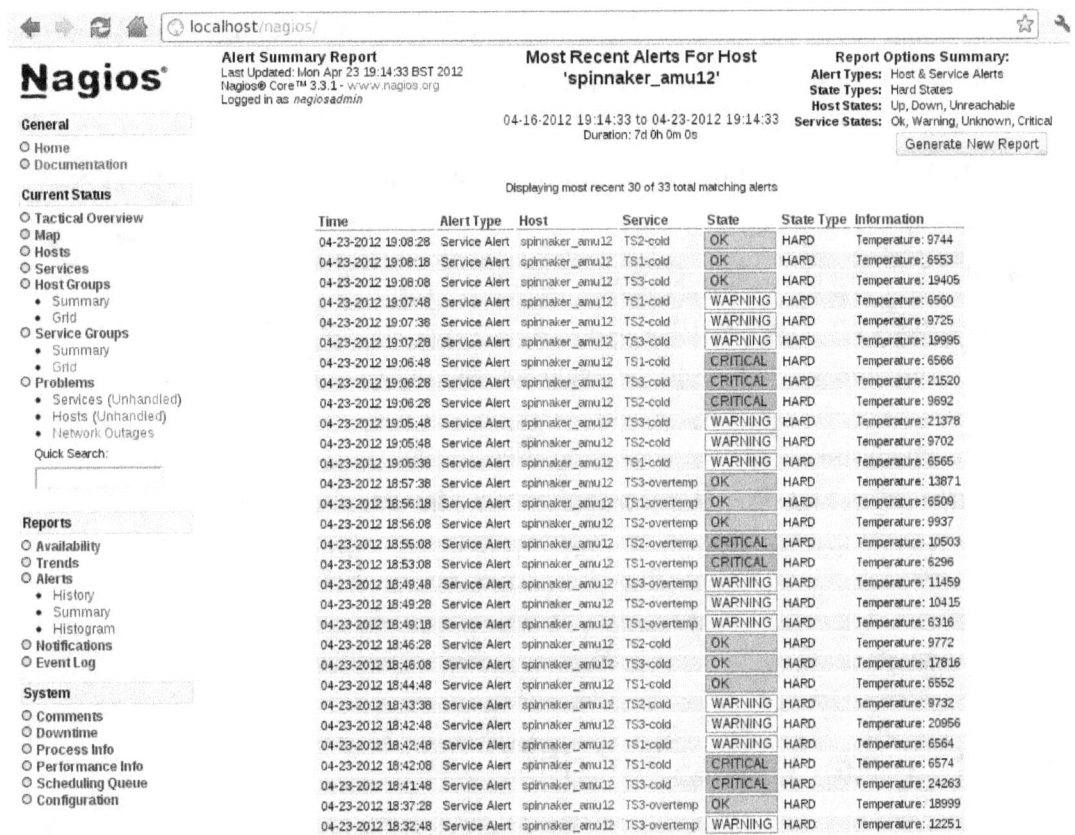

Figure 8.18: Alerting with Nagios – all 4 chips on a test board are 100% loaded (to warm them up), then pass into a quiescent mode where they cool down, both extremes generate warning and critical alerts as defined temperature thresholds are breached.

## 8.5 Monitoring a SpiNNaker Machine in Production

In the brain, no neuron shares a network connection, processing power or area to store input weightings, but on the SpiNNaker chip these facilities are contended artificially by resource sharing and rely on the silicon being faster than biology to meet their constraints. Some of the more important / key SpiNNaker items to monitor are:

- Network Connections. These need to be monitored for bottlenecks and capacity issues (the networks have been sized based on statistical data of real neuron firing-rates). The system also needs to be monitored for link failures and errors – sending this information back to the management station, so if a threshold is breached then a faulty link can be shut down.

- Processor. The algorithm used for neural modelling is well known eg. the Izhikevich model [JFW08], so monitoring should look for unexpectedly high processor utilisation. If demand overwhelms a processor then the real-time nature of the machine could be lost. Given that the information between neurons is encoded in the timing and ordering of spike arrival then this may prove detrimental or disastrous to an ANN simulation.

- Memory. The quantities and rates of DMA requests (and failures) can be read from the system.
- Plus Others. There are many other counters, parameters and health check registers in the SpiNNaker MIB tree which it would be useful to have in the managed system – to be able to both view and set for control purposes.

Management information can also be used to gain visibility of under-utilised resources that are available to take more load – a useful side-effect which can potentially be used to dynamically reallocate resources if required within the system.

## 8.6 Summary and Contributions

The SpiNNaker architecture is primarily aimed at real-time simulation, particularly of neural networks, and is formed of large numbers of components such as cores, chips, interconnects and memories. Having such a massively-parallel system operating brings with it challenges of how to manage its component health as it operates. System management facilities provide the required visibility, enabling hard and soft faults that occur to be managed in-flight by monitoring and alerting. Users (or automated agents) may then take appropriate actions – for example to re-route traffic or load onto alternative resources. This all must be undertaken in a lightweight fashion. Each SpiNNaker core (and chip) has very limited computational and memory resources available to it, therefore attempting to implement a conventional management tool chain would prove difficult, if not impossible.

A bespoke solution to the system management of the massively-parallel, but constrained, SpiNNaker architecture has therefore been created. The SpiNNaker framework employs trivial GET and SET commands to ensure management station to chip communications are as lightweight as possible. These primitive commands support retrieval and setting of status on the SpiNNaker nodes, for the price of a memory access (as the SpiNNaker systems incorporate their peripherals in their memory-mapped architecture).

## 8.6. SUMMARY AND CONTRIBUTIONS

The higher complexity of the native management protocols is abstracted away from SpiNNaker by the protocol translation function; this dynamically converts requests from management tools into the low-cost primitive operations, thus minimising memory and computational overhead on the SpiNNaker platform. The protocol translation is performed by SpiNNmate, a layered tool which on one side handles the primitive operations with SpiNNaker, and on the other external application protocols and tools. For system management purposes two protocol modules were developed, one bespoke to support bulk memory operations (loading and saving of data), the other to support the standard management protocol: SNMP. To support SNMP access a private enterprise MIB database of the SpiNNaker memory map and registers was created to suit the configuration of the architecture; this is reusable across the management domain as it provides an object oriented and hierarchical view of a SpiNNaker system.

Several experiments were performed with SNMP to validate the functionality (via the protocol translating SpiNNmate) using common industry management tools such as Nagios [NE12], Cacti [Cac12] and via generic scriptable SNMP operations. The more typical system 'Key Performance Indicators' were queried in the experiments, from processor utilisation to network loading, and others to read the more dynamic temperature sensors. To these management tools, the SpiNNmate translator looks like a standard SNMP agent, configured with the full MIB tree supporting the SpiNNaker hardware. To the SpiNNaker side the translator merely queries specific memory locations on the chips for specific data at low overall computational cost. The translator (via the database) takes care of mapping the raw memory locations of the machine to the MIB object structure used and onwards to the external protocol polling performed by the management system(s).

The major contributions made in this area are the creation of the management framework to manage a resource-constrained massively-parallel computing system, and ensuring that the burden placed on it is minimised while the functionality is not compromised. Underlying the operation of the management framework is the novel protocol translation function, and its conversions between the relatively heavyweight management tool protocols to the primitive memory-based operations on the SpiNNaker side. This extensible modular function can work equally as well with bespoke and standard protocols and provides a common method of remote access to the SpiNNaker system for management (and other) purposes. For scaling the system may be extended by adding new SpiNNmate instances to improve capacity and resiliently distribute the management load.

# Chapter 9

# Discussion and Conclusions

This thesis covers almost all areas of the full system management time-line of the SpiNNaker platform (fig. 9.1), from the time the power is applied until it is removed. At power up, the ROM Node-Boot software is retrieved and used to initialise and test the hardware of each processor and chip in the system. Any faulty components are disabled, and diagnostics recorded in an accessible record so that the status of a node may be recovered – even if that node has been shut down due to a system level fault. The Node-Boot software then enables the next phase of software to be efficiently flood-filled to the platform.

As SpiNNaker is primarily designed to run large spiking neural networks, the majority of code and data flood-filled is to support these large simulations. Providing visibility and analysis of Artificial Neural Networks (ANNs) during operation is provided

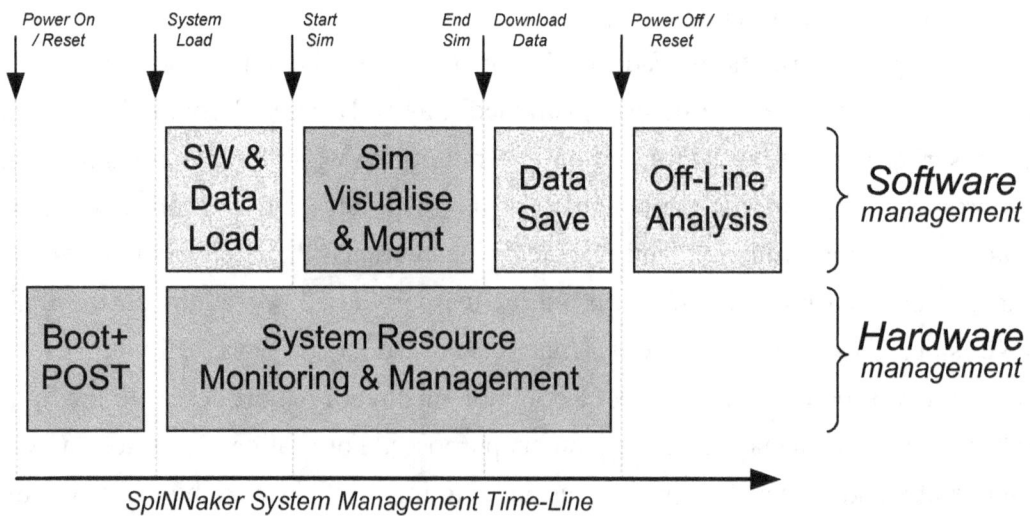

Figure 9.1: The system management time-line of the SpiNNaker platform.

for in the simulation visualisation software where user-selected components from the software model may be visualised as they run in real-time and interactions are performed via the same interface.

The SpiNNaker hardware, on which the simulations run, is a large distributed resource and its resources require management whilst the system is active and software executes. There are extensive ranges of items to monitor and adjust within a running system from hardware configuration to performance counters and gauges. A unified management framework has therefore been created to give 'low-cost' access to all levels of the machine. This access provides the required visibility to manage the machine as a whole, supporting access via bespoke and standard management tool sets.

The following sections discuss the research and development performed across all areas of the system management time-line, presented in the same order that the thesis has presented each area, and a section on the SpiNNaker platform itself.

## 9.1 Bootstrapping SpiNNaker

### 9.1.1 Primary Targets

The ROM image (or Node-Boot) is the first software which runs on all cores on a SpiNNaker chip after power on or reset. It performs Power-On Self-Tests (POST) and initialises the SpiNNaker node hardware into a state such that subsequent code may be loaded to it throughout an interconnected SpiNNaker system. If a fault is detected during POST then appropriate actions are taken to isolate the faulty component(s) where it is critical to a core or the system itself. This fault is recorded such that downstream management systems may determine the node-by-node health of the system, and use this information to aid in work allocation. The main goals were to ensure that the Node-Boot software is 'right-first-time' as the code is committed to silicon, and that it can reliably and repeatedly receive the next layer of software throughout the machine. If these two constraints are fulfilled, then any other functionality may then be loaded to the system subsequent to boot as software, whereas the ROM is immutable.

The Node-Boot software has been successful in its task of delivering these requirements for the SpiNNaker chips, the first of which were delivered from manufacture in May 2011, and subsequently no serious fault with Node-Boot has been detected in operation on the demonstration and test systems.

### 9.1.2 Time Taken to Boot

In reality SpiNNaker chips do not run Node-Boot code for long (hardware is passive and not productive in this mode), but it is the code which operates most frequently on the SpiNNaker platform as it runs on every core on every chip at every reset. Each node takes 4.0 seconds to enter its listening state after performing the POST and initialisation (predominantly TCM memory testing), and it then takes 5.3 seconds to distribute a 32 kB image by flood-fill to all nodes within a $2^8 \times 2^8$ maximum sized system; if performed consequently these operations total less than 10 seconds.

One way to judge the acceptability of this performance is to examine the overall end-to-end time of a simulation. Using the 4-chip SpiNNaker demonstration boards, a modest simulation may have its image flood-filled to the Application Processors and then the chips individually loaded with half their RAM capacity: $4 \times 64$ MB. It takes $\sim 20s$ to transmit this volume of data at (the Ethernet's) 100 Mb / s, so if the simulation ran just 1 second, and then results were recovered (e.g. half the loaded data post-simulation), the end-to-end simulation time exceeds 30 seconds. If systems with larger networks are considered (data volume and transmission time scaling linearly), the boot time (time in Node-Boot) tends to insignificance when compared with the simulation and data load / save stages. This process can be judged a success.

### 9.1.3 Power-On Self-Test

The POST operates on all chips as they are reset or powered on. For all the physical SpiNNaker chips tested, the POST has detected (and recorded) all core / peripheral faults with the exception of a small number of esoteric faults on a handful of cores which pass the ROM self-test routines (and manufacturing tests). Although it is not feasible to test absolutely every combination of state within such a complicated chip, the POST targets the die areas most critically affected by manufacturing defects. By area by core this is the TCM memory – as shown by the dark blocks adjacent to each core in the chip plot (fig. 4.3) – and testing each core's TCM memory consumes $\sim$50% of the POST initialisation time. If the core is sufficiently functional to execute and complete the RAM tests by (and the myriad of other tests it performs on its local peripherals), then there was an implicit assumption that the core would be fully functional.

This assumption turned out to be incorrect for the case of 1 core in the 3,000 or so tested so far (150+ SpiNNaker chips). The result of this is that in future all SpiNNaker chips, before and after packaging, will be tested by a more comprehensive test suite, to eliminate chips with cores which do not meet the 17-core yield constraints, or in particular: would not be detected by the POST testing and disabled. This section of the ROM therefore can be considered only as a partial success.

### 9.1.4 ITCM Validation Block (IVB)

The IVB enables users to break out of the normal boot sequence should a watchdog reset be triggered by code which is no longer responding as expected. Users create an image with a particular routine which is called in the event of a watchdog reset and may find it possible to recover to a 'safe' mode without losing all the state on the problematic node. The main issue around the resetting of a node is not necessarily the loss of a data on that node, but the impact it may have on the routing of data in a large machine, as routing tables are cleared at reset by Node-Boot.

The likely take-up of IVB is unknown at this stage and, while there is the advantage of being able to save data and the routing tables, it comes at the cost of the space required for the recovery routine within the ITCM image. It may be more likely that an IVB enabled image saves the routing tables if they appear credible, sacrificing the node's applications, and then awaits external assistance.

### 9.1.5 DHCP Node-Boot Image

The DHCP Node-Boot code is a variant of the ROM software which provides SpiNNaker chips the facility to retrieve a dynamically assigned IP address from a DHCP server managed on the local LAN. This is useful for users working in dynamically addressed networks (which are in the majority), as no static reconfiguration of the board is required when moving it between different LAN connections. The DHCP Node-Boot code has been successful tested in many LAN environments in conjunction with routines to auto-discover SpiNNaker boards to enable them to be located and used. There is a small cost to be paid in boot time which, as the image must be read from the external ROM chip rather than internally, extends initialisation from 4.0 to 5.5 seconds; there is then a further delay whilst a DHCP server is found and an IP address lease negotiated, which may be another few seconds. This disadvantage is generally outweighed by advantages of the additional DHCP functionality.

### 9.1.6 Future Work

Whilst the case for future work on a ROM which cannot be changed is slim as it is 'cast in silicon'; there are a small number of lessons learned and improvements that could be considered for a future edition of a SpiNNaker chip.

1. Implement electronic fuses which can be 'blown' to indicate that a core / chip should not be used. Whilst this would be useful for cores which have faults the POST cannot detect, it would also assist in cases where a production chip goes faulty in the same circumstances by disabling hardware before the POST is even performed.

2. There is a momentary current spike during boot as each processor switches from its 10 MHz boot to its Node-Boot 160 MHz operating frequency simultaneously. On a small-scale this is not problematic, but on larger systems, if this happens synchronously across dozens, hundreds or thousands of chips, the surge is significant. The resolution could have been to include a randomised back-off period before changing the clock rates. Furthermore, it was determined that this 'worst case' scenario may occur during SpiNNaker operation too, therefore early discovery has aided the power design of larger SpiNNaker systems.

3. Where more than one SpiNNaker chip has an Ethernet connection on the same local network, the 'Hello' messages from one chip can suppress the 'Hello' discovery messages from other chips. This is a mistake in the implementation which was intended to stop 'Hello' messages being transmitted once a Host system had started to communicate with that node (as it was already discovered). In the ROM code 'Hello' messages from other chips are erroneously interpreted as server messages, and further messages are quenched. This is not a critical issue, however, even when using auto-discovery, as a system may be booted through a single connection and one will always remain broadcasting. In any case, most SpiNNaker chip IP addresses are fixed and pre-populated into server software on the Host system.

4. Finally, if the POST is disabled by signalling the relevant GPIO pin (fig. 4.4), then not all status registers are reset (table 4.1) and may have indeterminate state. A workaround for validation is to check the GPIO pin to see if this is the case, but it would be preferable to have had the correct status recorded in these circumstances.

## 9.1. BOOTSTRAPPING SPINNAKER

Other than these points the ROM performs as designed and meets expectations. Overall the Node-Boot code installed on the ROM is a considered a success, as issues are non-critical or may be readily worked around.

**DHCP Discoverability**

The main 'problem' for the DHCP Node-Boot image on SpiNNaker remains its discoverability as the IP address provided will not be deterministic in most environments. Local auto-discovery is employed here, where the SpiNNaker chips send broadcast 'Hello' messages every few seconds. This technique is suitable where the Host is local to the chip (and broadcasts can be seen) and is a success in this environment.

However, if the Host is beyond the local LAN, the user will be blind to the board; this is a common problem where inter-network client-to-client connections are required. An example of such an application is Skype, where the problem is solved with a centralised 'directory' where both clients register to locate one another. A similar technique could be used for SpiNNaker with a central server coded into the SpiNNaker DHCP image; clients wishing to attach to a board would be presented with a list of SpiNNakers which have registered themselves with the directory service. Similarly, a Dynamic DNS (Domain Name System) client could be implemented where the SpiNNaker board reports to a central name server and is assigned a global DNS name which may be used to locate it. However, this only works where the chip is assigned a globally unique address and is attached to the Internet. A hybrid solution is furthermore being considered for the scenario where a static IP address is used in one place and a dynamic address elsewhere (e.g. office / home). Here the Node-Boot image would use a pre-programmed IP address unless a response is received from a DHCP server at reset.

Discovery problems are common with client devices across the Internet which run services and there is no perfect solution to the problem; combinations of the above techniques are used. Most typically, however, communications are client / server, with servers accessed by name, and mapped by DNS to a fixed IP address which changes infrequently. As larger installations of SpiNNaker boards become more prevalent the mobility of the SpiNNaker systems reduces, and the installation gravitates to client / server and fixed DNS type solutions for the SpiNNaker boards – which are more formally managed.

## 9.2 Visualising Neural Networks on SpiNNaker

As models of the brain become larger, more complex and run for longer periods, the ability to perform real-time data visualisation and analysis for artificial neural networks becomes more important. Aggregation of data is an inherent part of the visualisation process as it is feasible to plot only a subset of the data available in the system in real-time. If this aggregation can be achieved in parallel on the target architecture, then real-time visualisation and analysis can remain scalable. Viewing data from the simulations is achieved through a generalisable real-time visualisation tool.

### 9.2.1 Modularity

The purpose of the visualiser tool is to provide the type of visualisations found in both real and artificial neural networks and to display them, in real-time, along with the execution of the network. Visualisations are required to cover the full range of network behaviours. These span from plotting single neuron dynamics, neuron activity within a population, and population-based activity. These visualisations, too, emulate biological techniques from embedded single and arrays of electrodes, to EEG and fMRI. Many of the data plots used in neural network visualisations are standard graphing techniques and are used in multiple situations to represent various data sets. For this reason a generalised modular visualisation tool was created, which takes data from packets on a network, and interprets the data in the packet to populate data structures that may be plotted. This decision, not to plot the neural data directly, was taken to ensure that the tool is not tied to a particular simulator or packet format (it could be used with non-SpiNNaker neural simulators if required). Indeed, although its primary purpose is to plot neural data from SpiNNaker (and the modalities chosen are mostly aligned with this target), the visualiser may be used to plot any data that it receives, due to its modular nature. The visualiser approach has been a success and the results from chapter 6 cover a wide ranging set of real-time neural network visualisations from the SpiNNaker platform, plus non-neural network examples.

### 9.2.2 Modalities

As many of the same types of plot are used for neural network analysis, the tool has been created supporting many modalities that are suitable for neural data, e.g.: scatter plots (spike trains), line diagrams (parameter traces), tiled plots (aggregated activity maps / pseudo-fMRI), and histograms (synaptic weights). The benefit of these generalisable data plots is that context is abstracted and arbitrary, so the users may use any plot-type to view any type of data, and choose the appropriate colour-spectra, window sizes, transformations and 'zoom level' to display and interpret their real-time data. Many transformation and plot options, such as the multi-channel plot (where several channels of data may be plotted simultaneously over time), have been driven by user request. In this simulation the modular nature of the software proved useful, as the components from other plot types could be reused for items such as labelling of axes and scrolling of data.

The selection of C++ and OpenGL as a programming environment has been useful in the creation of windowed graphics that are portable across platforms, and in the efficient manipulation of data. The presentation of data on screen is accelerated by whatever graphics hardware is available on the machine; this has an impact on the capacity of the system. C++ has also been useful to create multiple threads to operate plotting and network modules semi-independently thus, ensuring non-blocking behaviour. At this stage, with small-scale networks, the visualisation modalities provided in the tool are successful in being able to represent the data required and, where necessary, it is easy to extend to add further visualisations, options or transformations.

### 9.2.3 Interaction

Interaction with simulations is the least developed function in the visualiser. The availability of the real-time visualisation has enabled the possibility of interacting with a running simulation, and this facility has been developed for a small number of networks. The majority of the interaction with simulations is provided through bespoke additions directly within the visualiser code. This is necessary as each interaction is simulation-specific as widely differing parameters may be in use, although some basic standardisation has been provided for in the high-level tools. For SpiNNaker ANNs options may be specified at a population level in the PyNN source code to enable specific 'reporting' to be turned on and off. An example of this is found in section 6.5.4, where the spike trains from a population are turned on and off via a single command sent to a specific population to 'zoom' in on its neurons' activity. By shifting focus a user then sends (implicit, by virtue of the code) instructions to turn off reporting that is out of scope. One section which has been standardised is the 'play, pause, stop' controls which output a standard SpiNNaker message to the simulation, which if set up to do so will react to the message.

Despite the standard interaction controls, as there is bespoke implementation required for each different visualisation interaction, this feature is therefore seen as a limited success. Future work is required in this area to develop a scalable approach to providing interaction with all simulation types.

### 9.2.4 Aggregation

Local aggregation is used to solve the problem of data collection and analysis in large platforms. In SpiNNaker each spike is represented as a 32-bit AER [LWM+93] message. A Gb / s (without overheads) is therefore required to convey 3 million neurons with a spike-rate of 10 Hz (before bifurcation). It is clearly infeasible to plot (or manually interpret) 30 million spikes on screen / s and the real-time visualisation scheme does not aim to offer this. Its philosophy is to aggregate the information internally, using spare 'utility' cores, to a representation which *is* practical to present and interpret on screen in real-time. In this example, the 3 million neurons may feasibly belong to ten thousand populations of 300 neurons each. The data can be plotted as a $100 \times 100$ grid, each tile representing the mean firing-rate of that particular population, which makes it feasible to present data from 3 million neurons in real-time. The neural mapping software is set up to add a local 'pseudo-neuron' to the output tree of all the neurons in each monitored population. This aggregator receives all spikes from a population and manipulates the information as required – in this example each local aggregator only has to count, on average, 3,000 spikes / s. Depending on how frequently the real-time visualisation is updated determines the rate of data sent from each aggregation point to the visualiser. In the example, if the update rate is 5 frames / s, there are $5 \times 10,000$ events / s, a feasible rate to plot and to interpret by eye (particularly if proximate population behaviour is closely aligned). This process may then be cascaded hierarchically to aggregate information further if required, thus reducing the quantity of information sent to the visualiser again.

This aggregated approach is feasible and was used in the section 6.5.4 experiment to provide the spike-rate display which may then be adjusted by interaction and zoomed into, spawning a new display. Aggregation also has a corollary with biological imaging, which represents *areas* of activity due to the limited resolution available to it. The full benefits of aggregation are yet to be experienced, as simulations with millions of neurons have not yet been built, but at this stage the example cited is a successful example of the aggregated data approach to real-time ANN visualisation.

### 9.2.5 Visualiser Limitations and Future Work

Due to limitations within the OpenGL environment for user-interaction, a rewrite of the code using Qt, wxWidgets or other such cross-platform framework which provides better handling of the user interface would be beneficial. Using a framework may provide a more structured approach than the current monolithic C++ implementation. The frameworks listed also support integration with OpenGL, so the application would not require complete redevelopment. A second possibility for the redevelopment of the visualiser is to use Python (as this language is commonly used in the ANN modelling community), but this would require more work than maintaining the use of C++. Some investigation as to the feasibility of using Python for this task would be required, as specific libraries and compilation may be required for performance purposes.

For future visualisation improvements it is envisaged to extend the real-time platform to add 3D projections, as currently only 2D in the form of $(x, y)$, or $(y, time)$ is supported. This should be relatively easily attainable as the OpenGL graphics engine natively supports 3D environments, but there will be some challenges around presenting and controlling this data from appropriate 'viewpoints' as data may be obscured. In the same vein, it is also intended to add a tomographic view to offer real-time slices through 3D data sets. This view can be extended to large-scale cortical simulations, such as [IE08] or the Blue Brain type activity visualisations [INC08, Con11]. Additionally SpiNNaker is being used for non-spiking neural simulation in large-scale multi-layer perceptrons (MLP) [JLP+10, RPWF12], therefore visualisation of weights and trajectories will be a target of these large simulations when operating.

The system is already able to make optional recordings of its data for off-line replay using the visualiser (and to replay at an arbitrary rate), with option to export this data in a format that may be plotted by Neurotools [DBE+09]. As the SpiNNaker machines scale in size it is likely that large, functional, cortical simulations will be executed and it may be useful to export data in a data format that medical imaging tools can parse [Gib08]. It is not the intention that the visualiser be used to save data for off-line analysis; network jitter may impact on its temporal accuracy and the aggregation may hide nuances in the data. Detailed analysis can, of course, be performed off-line by retrieving data from the machine itself after simulation.

Finally, the visualiser currently requires compilation for each visualisation type which sets up the parameters and interactions required to support that particular simulation. Two potential improvements to this situation are, firstly, to have the user select the visualisation required via a menu or the command line at start up; or secondly to take parameters from the network itself. Data about the network can be sourced from the mapping database [GDR+12] directly, or the simulation could transmit a specification to detail its visualisation and user-interaction requirements. Both these techniques would improve on the current hard-coded situation.

## 9.3 SpiNNaker Management Framework

SpiNNaker presents an unconventional route to providing high-performance parallel computing for neural computation by linking together large numbers of energy-efficient processors. The main architectural compromise is that each is resource-constrained compared with contemporary desktop or server processors. Operating in such a large distributed computing environment presents challenges to the retrieval of information from the machine as it operates. As there are so many components, it is essential to understand the health of the machine – with the hardware status guiding the placement of software elements on the system.

### 9.3.1 Protocol Translation

In chapter 8 the universal translator concept was introduced, which tackles the broad range of system communications demands on SpiNNaker, taking into account the limited resources available to each node. This function 'SpiNNmate' provides a simple low-overhead, but flexible and modular, approach to opening communication paths between two otherwise incompatible systems (fig. 8.3). The demands and complexity of external protocols are abstracted from the actual SpiNNaker nodes by SpiNNmate, ending up as a series of primitive operations which can be performed with minimal computational impact. Multiple SpiNNmate protocol handling modules provide support for diverse applications, with the core modules of SpiNNmate knitting together both sides of this translated conversation. External databases or other data sources may be introduced to aid in the effective translations between the systems.

Whilst the premise of the translation function clearly extends beyond just the hardware management domain, it was tested in this area. Two diverse management applications were created that cover bulk memory operations (MemGUI) and transactional retrieval of information (SNMP). In each of these cases the implementation was a success, covering the requirements for hardware management platforms for SpiNNaker using bespoke and standardised protocols.

## 9.3. SPINNAKER MANAGEMENT FRAMEWORK

### SpiNNaker Primitives

SpiNNaker's memory-mapped architecture means that the majority of peripherals, registers and memories can be accessed as locations in its memory map. Those peripherals and parameters which cannot be retrieved directly, may be accessed using sets of instructions executed locally on a processor, and message passing activities. The primitive operations are: 'GET', 'SET' and 'RUN'. The RUN command is necessarily more specialised it comprises a command code, rather than a memory address, and arguments which instruct the local SpiNNmate client code to perform one or a series of primitive operations in the system. The RUN command has been augmented with scheduler operations which enables local processing and aggregation of information to be performed in a distributed manner on board the SpiNNaker system. This avoids the necessity to request and send all raw data out of the system repeatedly, thus aiding management scaling, a similar philosophy to the aggregation of visualisation data in chapter 6. Scheduled operations may be layered from simple GET / SET instructions, for example to retrieve a parameter periodically and perform a rolling average calculation, or to test a parameter against a threshold and trigger an alert or change in state, all against a user defined schedule. Regardless of the external protocol translation enabled by SpiNNmate, the unified primitive commands are consistently used to provide the required interaction with the SpiNNaker system in real-time, and this scheme has been a success.

### MemGUI Protocol Handler

This GUI has been implemented and successfully provides bulk access to the memory map of a SpiNNaker chip, including integration with Host file operations. This implementation also successfully integrates with the database of SpiNNaker objects providing a more user-friendly method of accessing and influencing system state. While MemGUI is currently not the tool deployed to populate and recover memory blocks before and after simulations, it is a useful proof of concept implementation, and its GUI is more intuitive than the current command-line alternative, therefore succeeding in its purpose.

### SNMP Protocol Handler

SNMP is typically used for active periodic polling of system state across a number of performance variables of interest by a network management system. In chapter 8 a number of examples were deployed and the SNMP functionality to SpiNNaker has been proven. These examples used a number of industry standard SNMP tools and customised polling using SNMP to demonstrate the flexibility of the SpiNNaker implementation. The SNMP MIB design has been a success, distilling information on the SpiNNaker hardware configuration into a custom database and mapping onto its MIB tree. The querying of this database in real-time during SNMP object manipulation is also an implementation success, capable of over seven hundred SNMP polling operations / s, translating via SpiNNmate into the appropriate primitive commands. It is only when the truly large systems begin to be deployed that the benefit of using standardised tools for system management such as these will really show their benefits.

### 9.3.2 SpiNNmate Performance

The results of the functionality tests in chapter 8 have proved that the protocol translation concept is feasible; functionality is provided at low cost to the SpiNNaker system itself regardless of the external protocol complexity. The performance results achieved in the system are, unfortunately, limited due to the underlying SDP protocol whose overheads and performance are detailed in section 8.3.3. This ceiling in performance, together with the *current* restriction that the SpiNNmate responder cannot be incorporated into the Monitor Processor software (so must handle communications indirectly via message passing on a separate 'utility' core), results in performances in hundreds of kB / s, rather than in MB / s. Whilst creating the SpiNNmate primitive code, as well as testing the communication overheads (which are clearly the limiting factor on overall performance), the memory performance too has been assessed. This has been tested to ensure that the primitives use the most efficient method of accessing memory thus maximising the attainable memory bandwidth within the SpiNNaker chip itself. This optimisation work has led to new recommended defaults for memory transfer parameters; it has been applied to the SpiNNaker application framework and not just to SpiNNmate itself. These optimisations approximately triple the best performance previously possible in large block transfers; this is a success for SpiNNmate and also for understanding the SpiNNaker chip capabilities.

The poorest possible performance is encountered when variables are polled one memory location at a time and unfortunately this is how SNMP tends to issue its transactions. In the worst case an SNMP request for a single status bit may result in 128 bytes of network traffic in both the request and the reply. Clearly, performance can be improved by requesting blocks of data rather than individual items and by reducing the

## 9.3. SPINNAKER MANAGEMENT FRAMEWORK

frequency of polling; this is where the scheduled aggregation and transformation functions offered by RUN assist in reducing the burden. One other technique which was added to reduce overheads is that SpiNNmate primitives may be packed into a single packet. Here SpiNNmate may collate multiple requests and issue them simultaneously to the system to attain better efficiency of transmission. Ultimately performance is (and always will be) limited by network overheads, and this is discussed in section 9.4.

### 9.3.3 Future Work

The SpiNNmate system is currently used on small scale systems with a handful of chips and processors. As SpiNNaker systems are planned to increase in size tenfold with each deployment iteration, work is required to ensure monitoring via SpiNNmate does not become a bottleneck across its many components.

**Scaling for Future Requirements**

A good estimate of the total system management traffic required can be obtained by looking at a single node and extrapolating. Monitoring all 16 diagnostic counters, 8 DMA counters, processor utilisation, and the three temperature sensors of each chip would total 28 data items (words). Packing these into 4 primitives (same category counters are contiguous) requires 1 SDP packet with around 176 payload bytes. Monitoring these values every second, per node, can comfortably be accommodated by SpiNNmate's current performance and the current 1:4 ratio of Ethernets to SpiNNaker chips. Larger systems will have lower ratios of Ethernet:chip provision (potentially as low as 1:48), therefore, care is required in issuing management requests to limit the pace of resulting data. For management traffic, Ethernet-attached chips form the root (and potential hot-spot) of a management distribution tree. It remains an essential requirement that, during simulation, management traffic does not disrupt application traffic and, as all traffic is best effort with no service differentiation, this must be self-regulatory. A full 65,536 chip SpiNNaker configuration with the suggested monitoring requires ~10 MB / s. This requirement rises considerably during block memory operations (as the traffic is not throttled), therefore, in the future, it may be prudent to deploy multiple SpiNNmate instances per target system. Each SpiNNmate would use distributed Ethernets in the system to smooth demand and load on the SpiNNmate server itself.

### Improving SpiNNmate Performance

One simple way which may potentially improve the immediate performance of SpiNNmate is to permit more than 1 primitive packet to be in-flight at any one time. Currently, only a single request may be active until its acknowledgement is received (or it times out), so the network round-trip-time forms a restriction on the rate at which requests may be issued. This is currently being developed and tested, and should, in particular, improve the performance of wide-area transactions (where the latency is higher).

The network is currently the limiting factor on the performance of SpiNNmate, mainly due to the handling and transport of SDP packets, and their dissemination to the target SpiNNaker chip. One method to improve this is to interface directly with the SpiNNaker network itself and the opportunity to do this (using the board interconnection mechanisms of larger machines) is detailed in section 9.4 below.

### Placement of the Translator

SpiNNaker is its own massively-parallel computing resource. There is a potential avenue to explore on subsuming the SpiNNmate protocol translation functions within SpiNNaker itself; this is similar to the way the aggregation techniques of chapter 6 are performed internally. Rather than using external Host devices, Ethernet-attached SpiNNaker nodes not performing simulation work (utility chips), could be tasked with spreading the translation burden throughout the system. By using chips, rather than cores, the full 128 MB of SDRAM becomes available for SpiNNmate to work with; this is capacious enough to hold any data structures required (and has a further 16 or so cores to perform computational manipulations). These 'utility' chips would be responsible for sending out responses to target applications which are using their native protocol, and in this case the protocol translator would not require separate hardware.

### Extending the Scope

In this first implementation of SpiNNmate, the scope has been limited to the hardware function of the SpiNNaker machine; the intention, however, is to use this same mechanism to provide access to software structures internally in the system. SpiNNmate has been created with this link in mind, and other databases may be attached via the database layer (fig. 8.3). The link to software is facilitated by the database created when mapping neural networks to the target system architecture with PACMAN (section 3.5). By linking into this database, SpiNNaker primitives may be used to access software structures stored in local and system memories. Access via SNMP will be accommodated by generating SNMP MIBs systematically from the structures described in this database, enabling the same set of tools to support the management of both SpiNNaker hardware and its neural network software simulations.

Finally, there may be the potential to extend this low-overhead modular design to other target platforms by interchanging the SpiNNaker System Hardware layer within SpiNNmate (fig. 8.3). As outlined in section 1.1.1 the race is on, not just for the fastest computer, but for the most energy efficient high-performance computer, and these machines require management facilities too.

## 9.4 General SpiNNaker Observations

SpiNNaker chips have been available since May 2011 and have been used extensively since then for simulation and testing work. Simulations of a fraction of the full million+ processor system are currently being undertaken on small scale demonstrator boards which provide 72 processors over 4 chips (fig. 4.13). The size of SpiNNaker systems has risen substantially with the June 2012 delivery of modular 48-chip (864 processor) boards, and in the latter half of 2012 these boards will be interconnected to produce a $\sim$10,000 processor machine. The full-size million+ processor machines are targeted for construction and operation in 2013 and beyond.

Although the SpiNNaker chips appear fully functional, with no major faults, there are a couple of snags. Firstly the inter-chip links run at approximately one quarter of their intended rate, although the bandwidth they do provide remains well within the tolerances for simulating biologically realistic spiking neural networks [PGP+12]. Additionally, in power efficiency when a processor is asleep (low-power mode) its peripherals remain clocked, a design oversight, meaning the granularity of significant power saving is at a clocking domain level (9 processors), rather than individually.

Perhaps the biggest disappointment from a SpiNNaker management point of view arises from the software communications aspect of the machine. The SDP layer, which was developed to overlay the heterogeneous networks in the machine, is successfully in use to distribute code and data in the small SpiNNaker systems today; however its performance is constrained by the hardware processing it, and compounded by the low effective payload:header ratio for data transmission. For larger systems it appears these problems can be largely overcome...

### 9.4.1 Interconnecting Larger Machines

As has been identified, in transactional traffic, the network overhead created by SDP and Ethernet encapsulations is disproportionate and limits the total throughput of the system. In future large systems, modular 48-chip SpiNNaker circuit boards will be interconnected using fast serial connections, enabled by FPGAs, to multiplex interchip asynchronous links. This interposing of FPGA chips into the mesh will facilitate the direct low-latency insertion of packets into the system via the FPGA input / output channels. These paths may be taken directly to a Host device running SpiNNmate and thus instantly remove the overheads and requirements of the Ethernet path; potentially the FPGAs can offer larger bandwidth Gigabit Ethernet connections too for wide-area connections. This direct attachment to the mesh also offers a secondary benefit in that the onerous fragmentation and reassembly of SDP when bridging between external and internal SpiNNaker networks will no longer be required, or form a bottleneck, for communications. These future developments will clearly provide greater throughput for SpiNNmate primitives, and better overall SpiNNmate performance freed from the majority of the network limiting constraints.

### 9.4.2 Improving Load and Save Times – FR Packets for SDP

Within the data load and data save management stages (fig. 9.1) SDP is currently used as the transmission mechanism. Data is loaded sequentially to each chip in the system and similarly recovered at the end of the simulation if required. As seen in section 8.3.3, due to the small payload of each SpiNNaker packet and the SDP overheads, the internal efficiency of these transfers cannot exceed 31%. Once the bottleneck of the Ethernet connection is removed this internal transmission rate will then be the new limiting factor. It is therefore proposed that the fixed route (FR) packet type is brought into use for SDP and a dynamically created FR path defined across the network between ingress point and the target SpiNNaker chip. The route for the FR packets is implicit, therefore all traffic would flow down this path and have the full 64-bits of each packet available to it. Some headers would be required, but this method could more than double the transmission rate in the system for what is its most intensive data transmission period. Once the transmissions for that node have completed the FR path is simply reconfigured dynamically to point at the next target.

### 9.4.3 Further Works in Software

The most significant current obstacle to large-scale ANN modelling on SpiNNaker is the distribution of neural data in large networks. Placing-and-routing on a Host and subsequently downloading the data is adequate for small networks but, as the network grows, this leads to unacceptable Host compute and data transmission times, even with the potential improvements listed above. It is therefore essential to distribute this load into the SpiNNaker machine itself and take advantage of its massively-parallel processing to ease the place-and-route and data distribution burden.

The scalable SpiNNaker architecture is also garnering interest for use in applications beyond the purely spiking neural space. In addition to Izhikevich [JGP+10] and LIF [RGJF10] spiking models, Multi-Layer Perceptron (MLP) networks [JLP+10, RPWF12] have been built. Partner institutions are already taking advantage of SpiNNaker machines to run distributed ray-tracing applications, finite element simulations, and other applications where its efficient massive-parallelism is beneficial.

## 9.5 Summary

The SpiNNaker system management time-line has run its course... for this document at least. Its three major components have delivered:

Firstly, a functional ROM. Node-Boot software held here is able to detect hardware faults and keep them logged so that a management system can use this information to route application load around issues in the system. Node-Boot can gracefully manage faults which cause a watchdog reset and it provides a reliable mechanism for efficient receipt of flood-filled software and data throughout a SpiNNaker system.

Secondly, Visualisation software. This permits the flood-filled software to be monitored in real-time, across a variety of modes and at any level of neural dynamics. The visualiser handles on-board aggregation and transformation of data so that it does not swamp the network and gives the user the option to interact with the simulation to influence its behaviour and zoom in on the interesting details.

Finally, a low-cost, modular management framework. This translates protocols into low-overhead operations for SpiNNaker, facilitates easy loading and saving of data, and allows bespoke and standard SNMP tools to manage the target SpiNNaker system (with the potential to extend to the monitoring of SpiNNaker's software).

Implementation of the time-line has delivered an analysis and management framework for the real-time massively-parallel SpiNNaker neural architecture.

# Appendix A

# Expansion of Abbreviations

*Abbreviation   Expansion*

| | |
|---|---|
| ACK | Acknowledgement |
| ACM | Association for Computing Machinery |
| AER | Address Event Representation |
| AFNI | Analysis of Functional NeuroImages |
| AMBA | Advanced Microcontroller Bus Architecture |
| AMD | Advanced Micro Devices |
| ANN | Artificial Neural Network |
| API | Application Programming Interface |
| APT | Advanced Processor Technologies |
| ARP | Address Resolution Protocol |
| ASIC | Application Specific Integrated Circuit |
| BC | Broadcast |
| BCI | Brain Computer Interfacing |
| BIMPA | Biologically Inspired Massively Parallel Architectures |
| BIOS | Basic Input Output System |
| BMC | Baseboard Management Controller |
| BOLD | Blood Oxygen Level Dependent |
| BrainScaleS | Brain-inspired multiScale computation in neuromorphic hybrid Systems |
| CAT | Computed Axial Tomography |
| CCIE | Cisco Certified Internetworking Engineer |
| CIM | Common Information Model |
| CLI | Command Line Interface |
| cm | centimetre |
| CMIP | Common Management Information Protocol |
| CMP | Chip MultiProcessor |
| CPU | Central Processing Unit |
| CRC | Cyclic Redundancy Check |

| | |
|---|---|
| CT | Computed Tomography |
| CVS | Concurrent Versions System |
| DB | DataBase |
| DHCP | Dynamic Host Control Protocol |
| DICOM | Digital Imaging and COmmunications in Medicine |
| DMA | Direct Memory Access |
| DNS | Domain Name System |
| DTCM | Data Tightly-Coupled Memory |
| DTI | Diffusion Tensor Imaging |
| ECoG | ElectroCorticoGraphy |
| EEG | ElectroEncephaloGraphy |
| EM | Electro-Magnetic |
| ER | Emergency Routing |
| FACETS | Fast Analog Computing with Emergent Transient States |
| FCAPS | Fault, Configuration, Accounting, Performance and Security |
| fMRI | functional Magnetic Resonance Imaging |
| FPAA | Field-Programmable Analogue Array |
| FPGA | Field-Programmable Gate Array |
| FPNA | Field-Programmable Neural Array |
| fps | frames per second |
| FR | Fixed Route |
| FSL | Functional magnetic resonance imaging of the brain Software Library |
| GALS | Globally Asynchronous, Locally Synchronous |
| GB | GigaByte |
| Gb | Gigabit |
| GHz | GigaHertz |
| GENESIS | GEneral NEural SImulation System |
| GLUT | OpenGL Utility Toolkit |
| GNU | GNU's Not Unix! |
| GPGPU | General-Purpose computing on Graphics Processing Unit |
| GPIO | General-Purpose Input / Output |
| GPU | Graphics Processing Unit |
| GUI | Graphical User Interface |
| HICANN | High Input Count Analog Neural Network |
| HP | Hewlett-Packard company |
| HPC | High-Performance Computer / Computing |
| HTTP | HyperText Transfer Protocol |
| HTTPS | HyperText Transfer Protocol Secure |
| Hz | Hertz |
| I / O | Input / Output |
| IAB | Internet Activities Board |
| IBM | International Business Machines corporation |
| ICMP | Internet Control Message Protocol |
| ICONIP | International Conference on Neural Information Processing |

| | |
|---|---|
| ID | IDentification |
| iDRAC | Integrated Dell Remote Access Controller |
| IEC | International Electrotechnical Commission |
| IEEE | Institute of Electrical and Electronics Engineers |
| IET | Institution of Engineering and Technology |
| IETF | Internet Engineering Task Force |
| IJCNN | International Joint Conference on Neural Networks |
| ILM | Intermediate Level Manager |
| INIT | INITialising |
| IP | Internet Protocol |
| IPMI | Intelligent Platform Management Interface |
| IRQ | Interrupt ReQuest |
| ISO | International Organization for Standardization |
| ITCM | Instruction Tightly-Coupled Memory |
| IVB | ITCM Validation Block |
| JMX | Java Management eXtensions |
| JPSTH | Joint Post Stimulus Time Histogram |
| JVM | Java Virtual Machine |
| kB | kiloByte |
| kb | kilobit |
| KVM | Keyboard, Video, Mouse |
| kW | kiloWatt |
| LAN | Local Area Network |
| LIF | Leaky Integrate and Fire |
| LTD | Long Term Depression |
| LTP | Long Term Potentiation |
| MAC | Media Access Control |
| MATLAB | MATrix LABoratory |
| m | metre |
| MB | MegaByte |
| Mb | Megabit |
| MC | MultiCast |
| MEG | MagnetoEncephaloGraphy |
| MFLOPS | Mega FLoating-point Operations Per Second |
| MHz | MegaHertz |
| MIB | Management Information Base |
| MIPS | Millions of Instructions Per Second |
| MLP | Multi-Layer Perceptron |
| mm | millimetre |
| MoNETA | Modular Neural Exploring Traveling Agent |
| MP | Monitor Processor |
| MPI | Message Passing Interface |
| MPSoC | Multi-Processor System on Chip |
| MRI | Magnetic Resonance Imaging |

| | |
|---|---|
| MRTG | Multi Router Traffic Grapher |
| ms | millisecond |
| MW | MegaWatt |
| MWh | MegaWatt hour |
| NCS | NeoCortical Simulator |
| NEF | Neural Engineering Framework |
| NEST | NEural Simulation Tool |
| NIRS | Near Infra-Red Spectroscopy |
| NMS | Network Management System |
| NN | Nearest Neighbour |
| NoC | Network on Chip |
| NRZ | Non-Return-to-Zero |
| NRZI | Non-Return-to-Zero Inverted |
| OID | Object IDentifier |
| OOB | Out-Of-Band |
| OpenCL | Open Computing Language |
| OpenGL | Open Graphics Library |
| OSI | Open Systems Interconnection |
| OT | Optical Tomography |
| PACMAN | Partitioning And Configuration MANager |
| PC | Personal Computer |
| PCA | Principal Component Analysis |
| PCSIM | Parallel neural Circuit SIMulator |
| PDU | Protocol Data Unit |
| PEG | PneumoEncephaloGraphy |
| PET | Positron Emission Tomography |
| PHY | PHYsical layer |
| PLL | Phase-Locked Loop |
| POST | Power-On Self-Test |
| PowerPC | Performance Optimization With Enhanced RISC - Performance Computing |
| PRTG | Paessler Router Traffic Grapher |
| PSTH | Post Stimulus Time Histogram |
| PyNN | Python Neural Networks |
| RAM | Random Access Memory |
| RF | Radio Frequency |
| RFC | Request For Comments |
| RISC | Reduced Instruction Set Computing |
| RMON | Remote MONitoring |
| RMS | Resource Management System |
| ROM | Read-Only Memory |
| RRD | Round-Robin Database |
| RSA | Remote Supervisor Adapter |
| RTS | Real-Time System |
| RTZ | Return-To-Zero |

| | |
|---|---|
| RX | Receive |
| s | seconds |
| SATA | Serial AT Attachment |
| SC | System Controller |
| SDP | SpiNNaker Datagram Protocol |
| SDRAM | Synchronous Dynamic Random Access Memory |
| SLA | Service Level Agreement |
| SMI | Structure of Management Information |
| SNMP | Simple Network Management Protocol |
| SNN | Spiking Neural Network |
| SPECT | Single Photon Emission Computer Tomography |
| SPM | Statistical Parametric Mapping |
| SQL | Structured Query Language |
| SQUID | Superconducting QUantum Interference Device |
| SRAM | Static Random Access Memory |
| SSH | Secure SHell |
| STDP | Spike Timing Dependent Plasticity |
| SVN | SubVersioN |
| SyNAPSE | Systems of Neuromorphic Adaptive Plastic Scalable Electronics |
| TB | TeraByte |
| Tb | Terabit |
| TCM | Tightly-Coupled Memory |
| TFLOPS | Tera FLoating-point Operations Per Second |
| TIPS | Tera Instructions Per Second |
| TLU | Threshold Logic Unit |
| TX | Transmit |
| UC | UniCast |
| UDP | User Datagram Protocol |
| USB | Universal Serial Bus |
| VIC | Vector Interrupt Controller |
| VLAN | Virtual Local Area Network |
| VLSI | Very Large Scale Integration |
| VPN | Virtual Private Network |
| W | Watts |
| WBEM | Web-Based Enterprise Management |
| WCCI | World Congress on Computational Intelligence |
| WMI | Windows Management Instrumentation |
| XML | eXtensible Markup Language |
| XOR | eXclusive OR |

# Bibliography

[ABB+03]    G. Almasi, L. Bachega, R. Bellofatto, J. Brunheroto, C. Cascaval, J. Castaños, P. Crumley, C. Erway, J. Gagliano, D. Lieber, P. Mindlin, J. E. Moreira, R. K. Sahoo, A. Sanomiya, E. Schenfeld, R. Swetz, M. Bae, G. Laib, K. Ranganathan, Y. Aridor, T. Domany, Y. Gal, O. Goldshmidt, and E. Shmueli. System Management in the BlueGene/L Supercomputer. In *Proceedings of the 17th International Symposium on Parallel and Distributed Processing*, IPDPS '03, Washington, DC, USA, 2003. IEEE Computer Society.

[Abe82]     Moshe Abeles. *Local cortical circuits : an electrophysiological study / Moshe Abeles*. Springer-Verlag, Berlin ; New York, 1982.

[ABF+94]    K. Asanovic, J. Beck, J.A. Feldman, N. Morgan, and J. Wawrzynek. A supercomputer for neural computation. In *Proc. 1994 IEEE Intl. Conf. on Neural Networks (ICNN)*, volume 1, pages 5–9, 1994.

[AESM09]    Rajagopal Ananthanarayanan, Steven K Esser, Horst D Simon, and Dharmendra S Modha. The cat is out of the bag: cortical simulations with $10^9$ neurons, $10^{13}$ synapses. In *Proceedings of the Conference on High Performance Computing Networking, Storage and Analysis*, SC '09, pages 63:1–63:12, New York, NY, USA, 2009. ACM.

[Ale97]     S. Alexander. *RFC 2132 DHCP Options and BOOTP Vendor Extensions*. 1997. [Online; accessed 3-Jun-2012], URL: http://tools.ietf.org/html/rfc2132.

[Amd67]     Gene M. Amdahl. Validity of the single processor approach to achieving large scale computing capabilities. In *AFIPS '67 Spring: Proceedings of the April 18-20, 1967, Spring Joint Computer Conference*, pages 483–485, New York, NY, USA, 1967.

[AMG04]     R. S. Alves, C. C. Marquezan, and L. Z. Granville. Experiences in the implementation of an SNMP-based high performance cluster management system. In *Proc. Ninth International Symposium* on *Computers and Communications ISCC 2004*, volume 2, pages 1136–1141, June 2004.

[AN00] L. F. Abbott and S. B. Nelson. Synaptic plasticity: taming the beast. *Nature neuroscience*, 3 Suppl:1178–1183, November 2000.

[ARM11] ARM Ltd. *AMBA Open Specifications*. 2011. [Online; accessed 3-Jun-2012], URL: http://www.arm.com/products/system-ip/amba/amba-open-specifications.php.

[ARM12] ARM Ltd. *Company Profile - ARM*. 2012. [Online; accessed 3-Jun-2012], URL: http://www.arm.com/about/company-profile/index.php.

[Awi97] F. Awiszus. Spike train analysis. *J. Neurosci. Methods*, 74:155–166, Jun 1997.

[BAFL+11] William (Bill) Allcock, Evan Felix, Mike Lowe, Randal Rheinheimer, and Joshi Fullop. Challenges of HPC monitoring. In *High Performance Computing, Networking, Storage and Analysis (SC), 2011 International Conference for*, pages 1–6, Nov. 2011.

[BB07] J M Bower and D Beeman. GENESIS (simulation environment). *Scholarpedia*, 2(3), 2007.

[BCFG03] P. Bonetto, G. Comis, A.R. Formiconi, and M. Guarracino. A new approach to brain imaging, based on an open and distributed environment. In *Neural Engineering, 2003. Conference Proceedings. First International IEEE EMBS Conference on*, pages 521–524, March 2003.

[BDM04] Tom Binzegger, Rodney J Douglas, and Kevan A C Martin. A quantitative map of the circuit of cat primary visual cortex. *J Neurosci*, 24(39):8441–8453, September 2004.

[BDM07] T. Binzegger, R. J. Douglas, and K. A. Martin. Stereotypical bouton clustering of individual neurons in cat primary visual cortex. *J. Neurosci.*, 27(45):12242–12254, November 2007.

[BDM09] Tom Binzegger, Rodney J. Douglas, and Kevan A. Martin. Topology and dynamics of the canonical circuit of cat V1. *Neural Networks*, 22(8):1071–1078, October 2009.

[Bet12] NIMH Bethesda. *AFNI Analysis of Functional NeuroImages*. 6th March 2012. [Online; accessed 3-Jun-2012], URL: http://afni.nimh.nih.gov/afni.

[BF02] J. Bainbridge and S. Furber. Chain: a delay-insensitive chip area interconnect. *Micro, IEEE*, 22(5):16–23, Sep/Oct 2002.

[BHK+99] F. Beckmann, K. Heise, B. Klsch, U. Bonse, M.F. Rajewsky, M. Bartscher, and T. Biermann. Three-dimensional imaging of nerve tissue by X-ray phase-contrast microtomography. *Biophysical Journal*, 76(1):98–102, 1999.

[Bis96] Christopher M. Bishop. *Neural Networks for Pattern Recognition*. Oxford University Press, USA, 1st edition, January 1996.

[BK06] Fabrice Bernhard and Renaud Keriven. Spiking neurons on GPUs. In Vassil Alexandrov, Geert van Albada, Peter Sloot, and Jack Dongarra, editors, *Computational Science ICCS 2006*, volume 3994 of *Lecture Notes in Computer Science*, pages 236–243. Springer Berlin / Heidelberg, 2006.

[BKM+91] J. W. Belliveau, D. N. Kennedy, R. C. McKinstry, B. R. Buchbinder, R. M. Weisskoff, M. S. Cohen, J. M. Vevea, T. J. Brady, and B. R. Rosen. Functional mapping of the human visual cortex by magnetic resonance imaging. *Science*, 254:716–719, Nov 1991.

[BKM04] E. N. Brown, R. E. Kass, and P. P. Mitra. Multiple neural spike train data analysis: state-of-the-art and future challenges. *Nat. Neurosci.*, 7:456–461, May 2004.

[BL73] T. V. Bliss and T. Lomo. Long-lasting potentiation of synaptic transmission in the dentate area of the anaesthetized rabbit following stimulation of the perforant path. *J. Physiol. (Lond.)*, 232(2):331–356, Jul 1973.

[BMT09] Ivan Bogdanov, Radu Mirsu, and Virgil Tiponut. Matlab model for spiking neural networks. In *Proceedings of the 13th WSEAS international conference on Systems*, ICS'09, pages 533–537, Stevens Point, Wisconsin, USA, 2009. World Scientific and Engineering Academy and Society (WSEAS).

[Boa98] Kwabena Boahen. Communicating neuronal ensembles between neuromorphic chips. In *Neuromorphic Systems Engineering*, volume 447 of *The Kluwer International Series in Engineering and Computer Science*, pages 229–259. Springer US, 1998.

[Boa00] K.A. Boahen. Point-to-point connectivity between neuromorphic chips using address events. *Circuits and Systems II: Analog and Digital Signal Processing, IEEE Transactions on*, 47(5):416–434, May 2000.

[Boa06] K. Boahen. Neurogrid: Emulating a million neurons in the cortex. In *Engineering in Medicine and Biology Society, 2006. EMBS '06. 28th Annual International Conference of the IEEE*, Sept 2006.

[Bok81] S. H. Bokhari. On the mapping problem. *IEEE Trans. Comput.*, 30:207–214, March 1981.

[BPS10] M A Bhuiyan, V K Pallipuram, and M C Smith. Acceleration of spiking neural networks in emerging multi-core and GPU architectures. In *Parallel Distributed Processing, International Symposium on*, pages 1–8, 2010.

[Bra12a] Brain Computation Lab, University of Nevada, Reno. *Neurocortical Simulator (NCS)*. 2012. [Online; accessed 3-Jun-2012], URL: http://www.cse.unr.edu/brain/ncs.

[Bra12b] BrainScales. *Brain-inspired multiscale computation in neuromorphic hybrid systems*. 2012. [Online; accessed 3-Jun-2012], URL: http://brainscales.kip.uni-heidelberg.de/.

[BRC+07] Romain Brette, Michelle Rudolph, Ted Carnevale, Michael Hines, David Beeman, James Bower, Markus Diesmann, Abigail Morrison, Philip Goodman, Frederick Harris, Milind Zirpe, Thomas Natschlger, Dejan Pecevski, Bard Ermentrout, Mikael Djurfeldt, Anders Lansner, Olivier Rochel, Thierry Vieville, Eilif Muller, Andrew Davison, Sami El Boustani, and Alain Destexhe. Simulation of networks of spiking neurons: A review of tools and strategies. *Journal of Computational Neuroscience*, 23:349–398, 2007.

[Bro09] K. Brodmann. *Vergleichende Lokalisationslehre der Grosshirnrinde in ihren Prinzipien dargestellt auf Grund des Zellenbaues*. Johann Ambrosius Barth Verlag, Leipzig, 1909.

[BRP+10] A. Basu, S. Ramakrishnan, C. Petre, S. Koziol, S. Brink, and P.E. Hasler. Neural dynamics in reconfigurable silicon. *Biomedical Circuits and Systems, IEEE Transactions on*, 4(5):311–319, Oct. 2010.

[Cac12] Cacti Group, Inc. *Cacti - The Complete RRDTool-based Graphing Solution*. 2012. [Online; accessed 3-Jun-2012], URL: http://www.cacti.net/.

[CAG11] A. Cassidy, A.G. Andreou, and J. Georgiou. Design of a one million neuron single FPGA neuromorphic system for real-time multimodal scene analysis. In *Information Sciences and Systems (CISS), 2011 45th Annual Conference on*, pages 1–6, March 2011.

[Caj99] S.R. Cajal. *Comparative study of the sensory areas of the human cortex*. 1899.

[CECB03] E. Culurciello, R. Etienne-Cummings, and K.A. Boahen. A biomorphic digital image sensor. *Solid-State Circuits, IEEE Journal of*, 38(2):281–294, Feb 2003.

[Cer88] V. Cerf. *IAB Recommendations for the Development of Internet Network Management Standards*. 1988. [Online; accessed 3-Jun-2012], URL: http://tools.ietf.org/html/rfc1052.

[CfCNotDIfBB12]  Cognition Centre for Cognitive Neuroimaging of the Donders Institute for Brain and Behaviour. *Fieldtrip*. 2012. [Online; accessed 3-Jun-2012], URL: http://fieldtrip.fcdonders.nl/.

[CFE12]  CFEngine AS. *CFEngine - Distributed Configuration Management*. 2012. [Online; accessed 3-Jun-2012], URL: http:\www.cfengine.com.

[CFSD89]  J Case, M Fedor, M Schoffstall, and C Davin. *RFC 1067 A Simple Network Management Protocol (SNMP)*. 1989. [Online; accessed 3-Jun-2012], URL: http://tools.ietf.org/html/rfc1098.

[Cha84]  D. M. Chapiro. *Globally-asynchronous locally-synchronous systems*. Stanford University, CA, Ph.D. thesis, 1984.

[Cis07]  Cisco Systems. *Network Management System: Best Practices White Paper*. 2007. [Online; accessed 3-Jun-2012], URL: http://www.cisco.com/en/US/tech/tk869/tk769/technologies_white_paper09186a00800aea9c.shtml.

[Cis12a]  Cisco. *CiscoWorks LAN Management*. 2012. [Online; accessed 3-Jun-2012], URL: http://www.cisco.com/go/ciscoworks.

[Cis12b]  Cisco Systems. *Cisco Systems Certifications Homepage*. 2012. [Online; accessed 3-Jun-2012], URL: http://www.cisco.com/go/certifications.

[CMRW93]  J. Case, K. McCloghrie, M. Rose, and S. Waldbusser. *Structure of Management Information for version 2 of the Simple Network Management Protocol SNMPv2*. 1993. [Online; accessed 3-Jun-2012], URL: http://tools.ietf.org/html/rfc1442.

[Con11]  International Supercomputing Conference. *Henry Markram: Simulating the Brain The Next Decisive Years*. 2011. [Online; accessed 3-Jun-2012], URL: http://www.youtube.com/watch?v=h06lgyES6Oc.

[CS91]  Mark W. Craven and Jude W. Shavlik. Visualizing learning and computation in artificial neural networks. *International Journal on Artificial Intelligence Tools*, 1:399–425, 1991.

[CSSGSGLB07]  J. Costas-Santos, T. Serrano-Gotarredona, R. Serrano-Gotarredona, and B. Linares-Barranco. A spatial contrast retina with on-chip calibration for neuromorphic spike-based AER vision systems. *Circuits and Systems I: Regular Papers, IEEE Transactions on*, 54(7):1444–1458, July 2007.

[CTM+10]  Moran Cerf, Nikhil Thiruvengadam, Florian Mormann, Alexander Kraskov, Rodrigo Q. Quiroga, Christof Koch, and Itzhak Fried. On-line, voluntary control of human temporal lobe neurons. *Nature*, 467(7319), October 2010.

[DA01]	Peter Dayan and L. F. Abbott. *Theoretical Neuroscience: Computational and Mathematical Modeling of Neural Systems*. The MIT Press, 1st edition, 2001.

[Dan18]	W. E. Dandy. Ventriculography following the injection of air into the cerebral ventricles. *Ann Surg*, 68(1):5–11, Jul 1918.

[DB91]	Thomas A. Defanti and Maxine D. Brown. Advances in computers. In Marshall C. Yovits, editor, *Advances in computers*, chapter Visualization in scientific computing, pages 247–305. Academic Press Professional, Inc., San Diego, CA, USA, 1991.

[DBE+09]	Andrew P Davison, Daniel Brüderle, Jochen M Eppler, Jens Kremkow, Eilif Muller, Dejan Pecevski, Laurent Perrinet, and Pierre Yger. PyNN: a common interface for neuronal network simulators. *Frontiers in Neuroinformatics*, 2(0), 2009.

[Def12]	Defense Sciences Office. *Systems of Neuromorphic Adaptive Plastic Scalable Electronics (SyNAPSE)*. 2012. [Online; accessed 3-Jun-2012], URL: http://www.darpa.mil/Our_Work/DSO/Programs/Systems_of_Neuromorphic_Adaptive_Plastic_Scalable_Electronics_(SYNAPSE).aspx.

[Del12a]	Dell Inc. *Embedded Server Management*. 2012. [Online; accessed 3-Jun-2012], URL: http://content.dell.com/us/en/enterprise/dcsm-embedded-management.aspx.

[Del12b]	Dell Inc. *The Leading Systems Management Appliances - Dell KACE*. 2012. [Online; accessed 3-Jun-2012], URL: http://www.kace.com/.

[dGSGR10]	Hugo de Garis, Chen Shuo, Ben Goertzel, and Lian Ruiting. A world survey of artificial brain projects, Part I: Large-scale brain simulations. *Neurocomputing*, 74:3–29, 2010.

[DHMW08]	James Doyle, Matthew Holt, Cindy Mestad, and Steve Westerbeck. *Deployment Guide for Advanced Monitoring of a Blue Gene Environment*. January 2008. [Online; accessed 3-Jun-2012], URL: http://www.redbooks.ibm.com/abstracts/redp4356.html.

[Dim12]	Ivan Dimkovic. *SpikeFun*. 2012. [Online; accessed 3-Jun-2012], URL: http://www.dimkovic.com.

[Dis12]	Distributed Management Task Force Inc. *CIM — DMTF*. 2012. [Online; accessed 3-Jun-2012], URL: http://dmtf.org/standards/cim.

[DLP03]	Jack J. Dongarra, Piotr Luszczek, and Antoine Petitet. The LINPACK benchmark: past, present and future. *Concurrency and Computation: Practice and Experience*, 15(9):803–820, 2003.

[dLVA+98]  A. Van der Linden, M. Verhoye, J. Van Audekerke, R. Peeters, M. Eens, S. W. Newman, T. Smulders, J. Balthazart, and T. J. DeVoogd. Non invasive in vivo anatomical studies of the oscine brain by high resolution MRI microscopy. *Journal of Neuroscience Methods*, 81(1-2):45–52, 1998.

[DM91]  R. J. Douglas and K. A. Martin. A functional microcircuit for cat visual cortex. *J. Physiol. (Lond.)*, 440:735–769, 1991.

[DMTF10]  Inc. Distributed Management Task Force. *Web-Based Enterprise Management*. January 1999-2010. [Online; accessed 14-Jan-2012], URL: http://www.dmtf.org/standards/wbem.

[Dov12]  Dover Beach Consulting, Inc. *CMIP: Beginners Questions and Answers*. 2012. [Online; accessed 3-Jun-2012], URL: http://penta2.ufrgs.br/gereint/snmp3.htm.

[Dow01]  J. E. Dowling. *Neurons and Networks: An Introduction to Behavioral Neuroscience*. Harvard University Press, ISBN: 0674004620, 2nd revised edition, 2001.

[DPG+10]  S. Davies, C. Patterson, F. Galluppi, A.D. Rast, D. Lester, and S.B. Furber. Interfacing real-time spiking I/O with the SpiNNaker neuromimetic architecture. *Australian Journal of Intelligent Information Processing Systems*, 11(1), 2010.

[DRGF11]  Sergio Davies, Alexander D. Rast, Francesco Galluppi, and Steve Furber. A forecast-based biologically-plausible STDP learning rule. In *IJCNN 2011, International Joint Conference on Neural Networks*, pages 1810–1817, 2011.

[Dro97]  R. Droms. *RFC 2131 Dynamic Host Configuration Protocol*. 1997. [Online; accessed 3-Jun-2012], URL: http://tools.ietf.org/html/rfc2131.

[dS12]  Barcelona Supercomputing Center Centro Nacional de Supercomputacion. *The Mont-Blanc Project*. 2012. [Online; accessed 3-Jun-2012], URL: http://www.montblanc-project.eu/.

[DWEF00]  M Daniele, B Wijnen, M Ellison, and D Francisco. *RFC 2741 Agent Extensibility (AgentX) Protocol*. 2000. [Online; accessed 3-Jun-2012], URL: http://tools.ietf.org/html/rfc2741.

[DWH82]  R Devalois, E Williamyund, and N Hepler. The orientation and direction selectivity of cells in macaque visual cortex. *Vision Research*, 22(5):531–544, 1982.

[EA03]  C Eliasmith and C H Anderson. *Neural engineering: Computation, representation, and dynamics in neurobiological systems*. MIT Press, Cambridge, MA, 2003.

[EFEL05]   M. Amdeand A. Efthymiou, T. Felicijan, D. A. Edwards, and L. Lavagno. Asynchronous on-chip networks. *IEE Proceedings Computers and Digital Techniques*, 152:273–285, Mar 2005.

[FB09]   Steve Furber and Andrew Brown. Biologically-Inspired Massively-Parallel Architectures - Computing Beyond a Million Processors. In *Application of Concurrency to System Design, International Conference on*, pages 3–12, 2009.

[fBI12]   Athinoula A. Martinos Center for Biomedical Imaging. *FreeSurfer - set of automated tools for reconstructing cortical surface from structural MRI data*. 6th May 2012. [Online; accessed 3-Jun-2012], URL: http://surfer.nmr.mgh.harvard.edu/.

[FC12]   W Feng and K. Cameron. *The Green 500 List :: Environmentally Responsible Supercomputing*. Nov 2012. [Online; accessed 24-Jul-2012], URL: http://www.green500.org/.

[fCN12]   Swartz Center for Computational Neuroscience. *EEGLAB - Open Source Matlab Toolbox for Electrophysiological Research*. 6th May 2012. [Online; accessed 3-Jun-2012], URL: http://sccn.ucsd.edu/eeglab/.

[FGH06]   E. Farquhar, C. Gordon, and P. Hasler. A field programmable neural array. In *Circuits and Systems, 2006. ISCAS 2006. Proceedings. 2006 IEEE International Symposium on*, May 2006.

[FH08]   Kayvon Fatahalian and Mike Houston. A closer look at GPUs. *Commun. ACM*, 51:50–57, October 2008.

[fN12]   Wellcome Trust Centre for Neuroimaging. *Statistical Parametric Mapping (SPM)*. 6th May 2012. [Online; accessed 3-Jun-2012], URL: http://www.fil.ion.ucl.ac.uk/spm/.

[Fos03]   Ian Foster. *The Grid: A New Infrastructure for 21st Century Science*, pages 51–63. John Wiley & Sons, Ltd, 2003.

[Fou12]   Python Software Foundation. *Python Programming Language Official Website*. Python Software Foundation, 2012. [Online; accessed 3-Jun-2012], URL: http://www.python.org/.

[Fre06]   Free Software Foundation. *Concurrent Versions System*. 2006. [Online; accessed 3-Jun-2012], URL: http://www.nongnu.org/cvs/.

[FRSL09]   Andreas Fidjeland, Etienne B. Roesch, Murray Shanahan, and Wayne Luk. Nemo: A platform for neural modelling of spiking neurons using GPUs. In *ASAP*, pages 137–144. IEEE, 2009.

[FT06]   Steve Furber and Steve Temple. Neural systems engineering. *Journal of the Royal Society Interface*, 4:193–206, 2006.

[FTB06a] Steve Furber, Steve Temple, and Andrew Brown. High-performance computing for systems of spiking neurons. In *Grand Challenge 5: Architecture of Brain and Mind, AISB workshop on*, pages 29–36, 2006.

[FTB06b] Steve Furber, Steve Temple, and Andrew Brown. On-chip and inter-chip networks for modeling large-scale neural systems. In *Circuits and Systems, ISCAS 2006. 2006 IEEE International Symposium on*, 2006.

[GD03] M. Gallagher and T. Downs. Visualization of learning in multilayer perceptron networks using principal component analysis. *Systems, Man, and Cybernetics, Part B: Cybernetics, IEEE Transactions on*, 33(1):28–34, Feb 2003.

[GDF+12] Francesco Galluppi, Sergio Davies, Steve Furber, Terry Stewart, and Chris Eliasmith. Real Time On-Chip Implementation of Dynamical Systems with Spiking Neurons. In *Neural Networks (IJCNN), The 2012 International Joint Conference on*. IEEE, 2012.

[GDG07] O Gewaltig, Markus Diesmann, and Marc-Oliver Gewaltig. NEST (NEural Simulation Tool). *Scholarpedia*, 2(4), 2007.

[GDR+12] F. Galluppi, S. Davies, A.D. Rast, T. Sharp, L.A. Plana, and S.B. Furber. A Hierarchical Configuration System for a Massively Parallel Neural Hardware Platform. In *The 2012 ACM International Conference on Computer Frontiers*, 2012.

[Gee05] David Geer. Industry trends: Chip makers turn to multicore processors. *Computer*, 38(5):11–13, May 2005.

[Gib08] Bernard Gibaud. The DICOM standard: A brief overview. *Molecular Imaging: Computer Reconstruction and Practice*, pages 229–238, 2008.

[GJ79] Michael R Garey and David S Johnson. *Computers and Intractability: A Guide to the Theory of NP-Completeness*. W. H. Freeman & Co., New York, NY, USA, 1979.

[GLS+05] J. Geddes, S. Lloyd, A. Simpson, M. Rossor, N. Fox, D. Hill, J.V. Hajnal, S. Lawrie, A. McIntosh, E. Johnstone, J. Wardlaw, D. Perry, R. Procter, P. Bath, and E. Bullmore. Neurogrid: using grid technology to advance neuroscience. In *Computer-Based Medical Systems, 2005. Proceedings. 18th IEEE Symposium on*, pages 570–572, June 2005.

[Goo08] Dan Goodman. Brian: a simulator for spiking neural networks in Python. *Frontiers in Neuroinformatics*, 2, 2008.

[GRDF10]　　Francesco Galluppi, Alexander Rast, Sergio Davies, and Steve Furber. A general-purpose model translation system for a universal neural chip. ICONIP - 17th International Conference on Neural Information Processing, 2010.

[Gro12]　　GroundWork Inc. *GroundWork - The Open Platform for IT Monitoring*. 2012. [Online; accessed 3-Jun-2012], URL: http://www.gwos.com/.

[GTC+00]　　Dan Gunter, Brian Tierney, Brian Crowley, Mason Holding, and Jason Lee. Netlogger: A toolkit for distributed system performance analysis. In *Proceedings of the 8th International Symposium on Modeling, Analysis and Simulation of Computer and Telecommunication Systems*, MASCOTS '00, Washington, DC, USA, 2000. IEEE Computer Society.

[Ham92]　　M. S. Hamalainen. Magnetoencephalography: a tool for functional brain imaging. *Brain Topography*, 5:95–102, 1992.

[HC97]　　M L Hines and N T Carnevale. The NEURON Simulation Environment. *Neural Computation*, 9(6):1179–1209, August 1997.

[HDD11]　　John Hunter, Darren Dale, and Michael Droettboom. *Matplotlib: python plotting – Matplotlib v1.1.0*. 2011. [Online; accessed 3-Jun-2012], URL: http://matplotlib.sourceforge.net/.

[Heb49]　　D. O. Hebb. *The Organization of Behavior: A Neuropsychological Theory*. Psychology Press, 1949.

[Hew12a]　　Hewlett-Packard Development Company, L.P. *HP Cluster Management Utility*. 2012. [Online; accessed 3-Jun-2012], URL: http://h20311.www2.hp.com/HPC/cache/412128-0-0-0-121.html.

[Hew12b]　　Hewlett-Packard Development Company, L.P. *HP iLO Management Engine - Overview*. 2012. [Online; accessed 3-Jun-2012], URL: http://h18013.www1.hp.com/products/servers/management/remotemgmt.html.

[Hew12c]　　Hewlett-Packard Development Company, L.P. *HP Openview family*. 2012. [Online; accessed 3-Jun-2012], URL: http://www.openview.hp.com/.

[HH52]　　A L Hodgkin and A F Huxley. A quantitative description of membrane current and its application to conduction and excitation in nerve. *The Journal of Physiology*, 117(4):500–544, 1952.

[HH09]　　Suzana Herculano-Houzel. The human brain in numbers: a linearly scaled-up primate brain. *Front Hum Neurosci*, 3(0), 2009.

[HH11]　　Michael Hanke and Yaroslav O Halchenko. Neuroscience runs on GNU / Linux. *Frontiers in Neuroinformatics*, 5(00008), 2011.

[HKKS03] Matthias Hovestadt, Odej Kao, Axel Keller, and Achim Streit. Scheduling in HPC resource management systems: Queuing vs. planning. In Dror Feitelson, Larry Rudolph, and Uwe Schwiegelshohn, editors, *Job Scheduling Strategies for Parallel Processing*, volume 2862 of *Lecture Notes in Computer Science*, pages 1–20. Springer Berlin / Heidelberg, 2003.

[HKS11] Michael Hines, Sameer Kumar, and Felix Schurmann. Comparison of neuronal spike exchange methods on a Blue Gene/P supercomputer. *Frontiers in Computational Neuroscience*, 5(00049), 2011.

[HMH+08] J. Harkin, F. Morgan, S. Hall, P. Dudek, T. Dowrick, and L. McDaid. Reconfigurable platforms and the challenges for large-scale implementations of spiking neural networks. In *Field Programmable Logic and Applications, 2008. FPL 2008. International Conference on*, pages 483–486, Sept. 2008.

[HMR87] G. E. Hinton, J. L. McClelland, and D. E. Rumelhart. Distributed representations. In *Parallel Distributed Processing: Volume 1: Foundations*, pages 77–109. MIT Press, Cambridge, 1987.

[HN89] R. Hecht-Nielsen. Theory of the backpropagation neural network. In *Neural Networks, 1989. IJCNN., International Joint Conference on*, pages 593–605, 1989.

[HN90] Robert Hecht-Nielsen. *Neurocomputing*. Addison-Wesley, 1990.

[HT10] Bing Han and T.M. Taha. Neuromorphic models on a GPGPU cluster. In *Neural Networks (IJCNN), The 2010 International Joint Conference on*, July 2010.

[IBM12a] IBM. *IBM Cluster software: Cluster Systems Management (CSM)*. 2012. [Online; accessed 3-Jun-2012], URL: http://www-03.ibm.com/systems/software/csm/index.html.

[IBM12b] IBM. *IBM eServer xSeries and BladeCenter Server Management*. 2012. [Online; accessed 3-Jun-2012], URL: http://www.redbooks.ibm.com/Redbooks.nsf/RedbookAbstracts/sg246495.html.

[IBM12c] IBM. *Tivoli Netview*. 2012. [Online; accessed 3-Jun-2012], URL: http://www-01.ibm.com/software/tivoli/products/netview/.

[IBT10] IBTA - InfiniBand Trade Association. *About InfiniBand*. 2010. [Online; accessed 3-Jun-2012], URL: http://www.infinibandta.org/content/pages.php?pg=about_us_infiniband.

[IE08] Eugene M. Izhikevich and Gerald M. Edelman. Large-scale model of mammalian thalamocortical systems. *Proceedings of the National Academy of Sciences*, 105:3593–3598, 2008.

[IHS06]    V. Iyer, T. M. Hoogland, and P. Saggau. Fast functional imaging of single neurons using random-access multiphoton (RAMP) microscopy. *J. Neurophysiol.*, 95:535–545, Jan 2006.

[ILBH+11]    Giacomo Indiveri, Bernabe Linares-Barranco, Tara Julia Hamilton, Andre van Schaik, Ralph Etienne-Cummings, Tobi Delbruck, Shih-Chii Liu, Piotr Dudek, Philipp Hafliger, Sylvie Renaud, Johannes Schemmel, Gert Cauwenberghs, John Arthur, Kai Hynna, Fopefolu Folowosele, Sylvain Saighi, Teresa Serrano-Gotarredona, Jayawan Wijekoon, Yingxue Wang, and Kwabena Boahen. Neuromorphic silicon neuron circuits. *Frontiers in Neuroscience*, 5(00073), 2011.

[INC08]    INCF. *Henry Markram - The Blue Brain Project Keynote Lecture.* 2008. [Online; accessed 3-Jun-2012], URL: http://www.youtube.com/watch?v=8iDR8Z-e_GU.

[Ini12]    European Exascale Software Initiative. *Mont-Blanc - European scalable and power efficient HPC platform.* 6th May 2012. [Online; accessed 3-Jun-2012], URL: http://www.eesi-project.eu/media/BarcelonaConference/Day2/13-Mont-Blanc_Overview.pdf.

[Int09]    Intel and Dell and HP and NEC Corporation. *IPMI - Intelligent Platform Management Interface.* 2009. [Online; accessed 3-Jun-2012], URL: http://www.intel.com/design/servers/ipmi/.

[ISO89a]    ISO. *Common Management Information Protocol.* 1989. [Online; accessed 3-Jun-2012], URL: http://www.iso.org/iso/catalogue_detail.htm?csnumber=29698.

[ISO89b]    ISO. *Information processing systems – Open Systems Interconnection – Basic Reference Model – Part 4: Management framework.* 1989. [Online; accessed 3-Jun-2012], URL: http://www.iso.org/iso/catalogue_detail.htm?csnumber=14258.

[ITU08]    ITU. *X.680 : Information technology - Abstract Syntax Notation One ASN.1: Specification of basic notation.* 2008. [Online; accessed 3-Jun-2012], URL: http://www.itu.int/rec/T-REC-X.680/en.

[Izh03]    Eugene M. Izhikevich. Simple model of spiking neurons. *Neural Networks, IEEE Transactions on*, 14:1569–1572, 2003.

[Izh04]    E.M. Izhikevich. Which model to use for cortical spiking neurons? *Neural Networks, IEEE Transactions on*, 15(5):1063–1070, September 2004.

[Izh06]    Eugene M. Izhikevich. Polychronization: Computation with Spikes. *Neural Computation*, 18(2):245–282, 2006.

[JCSM79] R J Jaszczak, Lee-Tzuu Chang, N A Stein, and F E Moore. Wholebody single-photon emission computed tomography using dual, large-field-of-view scintillation cameras. *Physics in Medicine and Biology*, 24(6), 1979.

[JFW08] Xin Jin, S.B. Furber, and J.V. Woods. Efficient modelling of spiking neural networks on a scalable chip multiprocessor. In *Neural Networks, 2008. IJCNN 2008. IEEE International Joint Conference on*, pages 2812–2819, June 2008.

[JGP+10] Xin Jin, F. Galluppi, C. Patterson, A. Rast, S. Davies, S. Temple, and S. Furber. Algorithm and software for simulation of spiking neural networks on the multi-chip SpiNNaker system. In *IJCNN 2010, International Joint Conference on Neural Networks*, pages 1–8, July 2010.

[JL07] C Johansson and A Lansner. Towards Cortex Sized Artificial Neural Systems. *Neural Networks*, 20(1):48–61, January 2007.

[JLP+10] Xin Jin, Mikel Luján, Luis A. Plana, Alexander D. Rast, Stephen R. Welbourne, and Steve B. Furber. Efficient parallel implementation of multilayer backpropagation networks on SpiNNaker. In *Proceedings of the 7th ACM International Conference on Computing Frontiers*, pages 89–90, New York, USA, 2010. ACM.

[Job77] F. F. Jobsis. Noninvasive, infrared monitoring of cerebral and myocardial oxygen sufficiency and circulatory parameters. *Science*, 198:1264–1267, Dec 1977.

[Jol02] I.T. Jolliffe. *Principal Component Analysis*. Springer Series in Statistics, 2002.

[JUL12] JULICH SUPERCOMPUTING CENTRE (JSC). *FZJ-JSC IBM Blue Gene/P - JUGENE home page*. 2012. [Online; accessed 3-Jun-2012], URL: http://www2.fz-juelich.de/jsc/jugene.

[Kas91] F. Kastenholz. *SNMP Communications Services*. 1991. [Online; accessed 3-Jun-2012], URL: http://tools.ietf.org/html/rfc1270.

[KBM02] Klaus Krauter, Rajkumar Buyya, and Muthucumaru Maheswaran. A taxonomy and survey of grid resource management systems for distributed computing. *Softw. Pract. Exper.*, 32(2):135–164, February 2002.

[KHH+09] James G King, Michael Hines, Sean L Hill, Philip H Goodman, Henry Markram, and Felix Schurmann. A component-based extension framework for large-scale parallel simulations in neuron. *Frontiers in Neuroinformatics*, 3(00010), 2009.

[KJOA07]  M. L. Kringelbach, N. Jenkinson, S. L. Owen, and T. Z. Aziz. Translational principles of deep brain stimulation. *Nat. Rev. Neurosci.*, 8:623–635, Aug 2007.

[Kot07]  Douglas B. Kothe. *Science Prospects and Benefits with Exascale Computing*. 2007. [Online; accessed 3-Jun-2012], URL: http://www.nccs.gov/wp-content/media/nccs_reports/Science%20Case%20_012808%20v3_final.pdf.

[KPP09]  Andrzej Kasinski, Juliusz Pawlowski, and Filip Ponulak. SNN3DViewer - 3D visualization tool for spiking neural network analysis. In *Computer Vision and Graphics*, volume 5337 of *Lecture Notes in Computer Science*, pages 469–476. Springer Berlin / Heidelberg, 2009.

[KRN+11]  M.M. Khan, AD Rast, J. Navaridas, X. Jin, L.A. Plana, M. Luján, S. Temple, C. Patterson, D. Richards, JV Woods, J. Miguel-Alonso, and S.B. Furber. Event-driven configuration of a neural network CMP system over an homogeneous interconnect fabric. *Parallel Computing*, 2011.

[KW96]  Sami Khuri and Jason Williams. Neuralis: an artificial neural network package. In *ITiCSE*, pages 25–27, 1996.

[KWC+09]  Sung-Su Kim, Y.J. Won, Mi-Jung Choi, J.W.-K. Hong, and J. Strassner. Towards management of the future internet. In *Integrated Network Management-Workshops, 2009. IM '09. IFIP/IEEE International Symposium on*, pages 81–86, June 2009.

[Lau73]  P. C. Lauterbur. Image formation by induced local interactions: Examples employing nuclear magnetic resonance. *Nature*, 242(5394):190–191, March 1973.

[LKC+10]  Victor W. Lee, Changkyu Kim, Jatin Chhugani, Michael Deisher, Daehyun Kim, Anthony D. Nguyen, Nadathur Satish, Mikhail Smelyanskiy, Srinivas Chennupaty, Per Hammarlund, Ronak Singhal, and Pradeep Dubey. Debunking the 100x GPU vs. CPU myth: an evaluation of throughput computing on CPU and GPU. *SIGARCH Comput. Archit. News*, 38(3):451–460, June 2010.

[LLM88]  Michael Litzkow, Miron Livny, and Matthew Mutka. Condor - a hunter of idle workstations. In *Proceedings of the 8th International Conference of Distributed Computing Systems*, June 1988.

[LPD08]  P. Lichtsteiner, C. Posch, and T. Delbruck. A $128 \times 128$ 120db 15us latency asynchronous temporal contrast vision sensor. *Solid-State Circuits, IEEE Journal of*, 43(2):566–576, Feb. 2008.

[LW09]  Jihong Liu and Chengyuan Wang. A survey of neuromorphic engineering–biological nervous systems realized on silicon. In

*Testing and Diagnosis, 2009. ICTD 2009. IEEE Circuits and Systems International Conference on*, pages 1–4, April 2009.

[LWM+93] J Lazzaro, J Wawrzynek, M Mahowald, M Silviotti, and D Gillespie. Silicon Auditory Processors as Computer Peripherals. *IEEE Transactions on Neural Networks*, 4(3):523–528, May 1993.

[Maa96] Wolfgang Maass. Networks of spiking neurons: The third generation of neural network models. *Neural Networks*, 10:1659–1671, 1996.

[Maa97] W. Maass. Fast sigmoidal networks via spiking neurons. *Neural Computation*, 9(2):279–304, 1997.

[Mar06] Henry Markram. The Blue Brain Project. *Nat. Rev. Neurosci.*, 7:153–160, 2006.

[MAS11] Jit Muthuswamy, Sindhu Anand, and Arati Sridharan. Adaptive movable neural interfaces for monitoring single neurons in the brain. *Frontiers in Neuroscience*, 5(00094), 2011.

[MASB07] Paul A. Merolla, John V. Arthur, Bertram E. Shi, and Kwabena A. Boahen. Expandable networks for neuromorphic chips. *Circuits and Systems I: Regular Papers, IEEE Transactions on*, 54(2):301–311, Feb. 2007.

[Mat12a] MathWorks. *MATLAB - The Language of Technical Computing*. 2012. [Online; accessed 3-Jun-2012], URL: http://www.mathworks.co.uk/products/matlab/.

[Mat12b] MathWorks. *MATLAB Central*. 2012. [Online; accessed 3-Jun-2012], URL: http://www.mathworks.co.uk/matlabcentral/.

[Mat12c] MathWorks. *Neural Network Toolbox - MATLAB*. 6th May 2012. [Online; accessed 3-Jun-2012], URL: http://www.mathworks.co.uk/products/neural-network.

[Mea89] C. A. Mead. *Analog VLSI and Neural Systems*. Addison-Wesley, 1989.

[Mea90] C. Mead. Neuromorphic electronic systems. *Proceedings of the IEEE*, 78(10):1629–1636, Oct 1990.

[MFM+12] Simon Moore, Paul Fox, Steven Marsh, A. Theodore Markettos, and Alan Mujumdar. Bluehive - a field-programable custom computing machine for extreme-scale real-time neural network simulation. In *Proceedings of the 2012 IEEE 20th Annual International Symposium on Field-Programmable Custom Computing Machines*, FCCM '12. IEEE Computer Society, 2012.

[Mic09] MicroDigital. *TCP/IP Stack Embedded Ethernet ARM*. 30th August 2009. [Online; accessed 3-Jun-2012], URL: http://www.smxrtos.com/tcpip.htm.

[Mou78]     V. Mountcastle. An organizing principle for cerebral function: the unit model and the distributed system. In *The Mindful Brain*. MIT Press, Cambridge, Mass., 1978.

[MP43]      Warren McCulloch and Walter Pitts. A logical calculus of the ideas immanent in nervous activity. *Bulletin of Mathematical Biology*, 5:115–133, 1943.

[MP69]      Marvin L. Minsky and Seymour A. Papert. *Perceptrons*. The MIT Press, December 1969.

[MPI09]     MPI Forum. *Message Passing Interface (MPI) Forum Home Page*. 2009. [Online; accessed 3-Jun-2012], URL: http://www.mpi-forum.org/.

[MPS+99]    K. McCloghrie, D. Perkins, J. Schoenwaelder, TU Braunschweig, J. Case, M. Rose, and S. Waldbusser. *Structure of Management Information Version 2 $SMIv2$*. 1999. [Online; accessed 3-Jun-2012], URL: http://tools.ietf.org/html/rfc2578.

[MR88a]     K. McCloghrie and M. Rose. *Management Information Base for Network Management of TCP/IP-based Internets*. 1988. [Online; accessed 3-Jun-2012], URL: http://tools.ietf.org/html/rfc1066.

[MR88b]     K. McCloghrie and M. Rose. *Structure and Identification of Management Information for TCP/IP-based Internets*. 1988. [Online; accessed 3-Jun-2012], URL: http://tools.ietf.org/html/rfc1065.

[MSDS12]    H. Meuer, E. Strohmaier, J. Dongarra, and H. Simon. *Top500 Supercomputing Sites*. Nov 2012. [Online; accessed 24-July-2012], URL: http://www.top500.org/.

[Mun92]     P.W. Munro. Visualizations of 2-d hidden unit space. In *Neural Networks, 1992. IJCNN., International Joint Conference on*, volume 3, pages 468–473, Jun 1992.

[MV94]      A. Mortara and E. A. Vittoz. A communication architecture tailored for analog VLSI artificial neural networks: intrinsic performance and limitations. *IEEE Transactions on Neural Networks*, 5:459–466, 1994.

[NAS11]     NASA. *NASA High-End Computing Program*. 2011. [Online; accessed 3-Jun-2012], URL: http://www.hec.nasa.gov/news/features/2011/sbus_042111.html.

[NDK+09]    Jayram Moorkanikara Nageswaran, Nikil D. Dutt, Jeffrey L. Krichmar, Alex Nicolau, and Alexander V. Veidenbaum. A configurable simulation environment for the efficient simulation of large-scale spiking neural networks on graphics processors. *Neural Networks*, pages 791–800, 2009.

[NE12]	LLC Nagios Enterprises. *Nagios - The Industry Standard In IT Infrastructure Monitoring*. 2012. [Online; accessed 3-Jun-2012], URL: http://www.nagios.org/.

[Neu11]	NeuroDebian. *Supplementary survey results – Debian Neuroscience Package Repository*. 2011. [Online; accessed 3-Jun-2012], URL: http://neuro.debian.net/survey/2011/results.html.

[NLMA+09]	Javier Navaridas, Mikel Luján, Jose Miguel-Alonso, Luis A. Plana, and Steve Furber. Understanding the Interconnection Network of SpiNNaker. In *Proceedings of the 23rd International Conference on Supercomputing*, ICS '09, pages 286–295, New York, NY, USA, 2009. ACM.

[NPMA+10]	Javier Navaridas, Luis A. Plana, Jose Miguel-Alonso, Mikel Luján, and Steve B. Furber. SpiNNaker: impact of traffic locality, causality and burstiness on the performance of the interconnection network. In *Proceedings of the 7th ACM international conference on Computing frontiers*, CF '10, pages 11–20, New York, NY, USA, 2010. ACM.

[NVI12]	NVIDIA. *NVIDIA TEGRA*. 6th May 2012. [Online; accessed 3-Jun-2012], URL: http://www.nvidia.com/object/tegra.html.

[Oet11]	Tobi Oetiker. *MRTG - The Multi Router Traffic Grapher*. 2011. [Online; accessed 3-Jun-2012], URL: http://oss.oetiker.ch/mrtg/.

[oFOU12]	Analysis Group of FMRIB Oxford UK. *FMRIB Software Library (FSL)*. 6th May 2012. [Online; accessed 3-Jun-2012], URL: http://www.fmrib.ox.ac.uk/fsl/.

[OG11]	The Open Group. *OpenPegasus*. 2011. [Online; accessed 3-Jun-2012], URL: http://www3.opengroup.org/subjectareas/management/openpegasus.

[OLKT90]	S. Ogawa, T. M. Lee, A. R. Kay, and D. W. Tank. Brain magnetic resonance imaging with contrast dependent on blood oxygenation. *Proc Natl Acad Sci U S A*, 87(24):9868–9872, December 1990.

[ONH+96]	Kunle Olukotun, Basem A. Nayfeh, Lance Hammond, Ken Wilson, and Kunyung Chang. The case for a single-chip multiprocessor. In *Proceedings of the seventh international conference on Architectural support for programming languages and operating systems*, ASPLOS-VII, pages 2–11, New York, NY, USA, 1996. ACM.

[Onl12]	Encyclopaedia Britannica Online. *brain: functional areas. [Art]*. 2012. [Online; accessed 3-Jun-2012], URL: http://www.britannica.com/EBchecked/media/100577/Functional-areas-of-the-human-brain.

[Ope06] OpenWBEM. *OpenWBEM*. 2006. [Online; accessed 3-Jun-2012], URL: http://openwbem.sourceforge.net/.

[Ope12] OpenNMS Group. *The OpenNMS Project*. 2012. [Online; accessed 3-Jun-2012], URL: http://www.opennms.org/.

[Ora08] Oracle. *Java Management Extensions (JMX)*. 2008. [Online; accessed 3-Jun-2012]. URL: http://www.oracle.com/technetwork/java/javase/tech/javamanagement-140525.html.

[PA97] P. Pfaerber and K. Asanovic. Parallel neural network training on MultiSpert. In *Proc. IEEE $3^{rd}$ Int'l Conf. on Algorithms and Architectures for Parallel Processing (ICA3PP)*, pages 659–666, Dec 1997.

[Pae12] Paessler AG. *PRTG Network Monitor - intuitive network monitoring software*. 2012. [Online; accessed 3-Jun-2012], URL: http://www.paessler.com/prtg.

[PBF+08] L.A. Plana, J. Bainbridge, S. Furber, S. Salisbury, Yebin Shi, and Jian Wu. An on-chip and inter-chip communications network for the SpiNNaker massively-parallel neural net simulator. In *Networks-on-Chip, 2008. NoCS 2008. Second ACM/IEEE International Symposium on*, pages 215–216, April 2008.

[PBS11] V.K. Pallipuram, M.A. Bhuiyan, and M.C. Smith. Evaluation of GPU architectures using spiking neural networks. In *Application Accelerators in High-Performance Computing (SAAHPC), 2011 Symposium on*, pages 93–102, July 2011.

[PEM+07] H. Plesser, J. Eppler, A. Morrison, M. Diesmann, and M.O. Gewaltig. Efficient parallel simulation of large-scale neuronal networks on clusters of multiprocessor computers. *Euro-Par 2007 parallel processing*, pages 672–681, 2007.

[PFT+07] L.A. Plana, S.B. Furber, S. Temple, M. Khan, Y. Shi, J. Wu, and S. Yang. A GALS Infrastructure for a Massively Parallel Multiprocessor. *Design & Test of Computers, IEEE*, 24(5):454–463, 2007.

[PGP+12] Cameron Patterson, Jim Garside, Eustace Painkras, Steve Temple, Luis A. Plana, Javier Navaridas, Thomas Sharp, and Steve Furber. Scalable communications for a million-core neural processing architecture. *Journal of Parallel and Distributed Computing (in press)*, 2012. DOI: 10.1016/j.jpdc.2012.01.016, http://www.sciencedirect.com/science/article/pii/S0743731512000287.

[PGRF12] Cameron Patterson, Francesco Galluppi, Alexander Rast, and Steve Furber. Visualising Large-Scale Neural Network Models in Real-Time. In *Neural Networks (IJCNN), The 2012 International Joint Conference on*. IEEE, 2012.

[Pla12]  Platform Computing and IBM Company. *Clusters, Grids, Clouds - Platform Computing*. 2012. [Online; accessed 3-Jun-2012], URL: http://www.platform.com/.

[PMKK91]  Lorien Y. Pratt, Jack Mostow, Candace A. Kamm, and Ace A. Kamm. Direct transfer of learned information among neural networks. In *Proceedings of AAAI-91*, pages 584–589, 1991.

[PN09]  D. Pecevski and T. Natschlager. *PCSIM: A Parallel neural Circuit SIMulator*. 2009. [Online; accessed 3-Jun-2012], URL: http://www.lsm.tugraz.at/pcsim/.

[PPG+12]  Eustace Painkras, Luis A. Plana, Jim Garside, Steve Temple, Simon Davidson, Jeffrey Pepper, David Clark, Cameron Patterson, and Steve Furber. SpiNNaker: A Multi-Core System-on-Chip for Massively-Parallel Neural Net Simulation. In *(accepted for) Custom Integrated Circuits Conference (CICC), 2012 IEEE*, 2012.

[PPGF12]  Cameron Patterson, Thomas Preston, Francesco Galluppi, and Steve Furber. Managing a massively-parallel resource-constrained computing architecture. In *(accepted for) 15th Euromicro Conference on Digital System Design*, 2012.

[PPM+07]  M.J. Pearson, A.G. Pipe, B. Mitchinson, K. Gurney, C. Melhuish, I. Gilhespy, and M. Nibouche. Implementing spiking neural networks for real-time signal-processing and control applications: A model-validated FPGA approach. *Neural Networks, IEEE Transactions on*, 18(5):1472–1487, Sept. 2007.

[Pre02]  R. Presuhn. *Management Information Base (MIB) for the Simple Network Management Protocol (SNMP)*. 2002. [Online; accessed 3-Jun-2012]. URL: http://tools.ietf.org/html/rfc3418.

[Que12]  Quest Software Inc. *Server monitoring and System management from Big Brother Software*. 2012. [Online; accessed 3-Jun-2012], URL: http://bb4.com/.

[Res58]  F. Resenblatt. The perceptron: a probabilistic model for information storage and organization in the brain. *Psychol Rev*, 65(6):386–408, Nov 1958.

[RGD+11]  Alexander Rast, Francesco Galluppi, Sergio Davies, Luis Plana, Cameron Patterson, Thomas Sharp, David Lester, and Steve Furber. Concurrent heterogeneous neural model simulation on real-time neuromimetic hardware. *Neural Networks*, 24(9):961–978, 2011.

[RGJF10]  A.D. Rast, F. Galluppi, X. Jin, and S.B. Furber. The Leaky Integrate-and-Fire neuron: A platform for synaptic model exploration on the SpiNNaker chip. In *IJCNN 2010, International Joint Conference on Neural Networks*, pages 1–8, July 2010.

[RJG+10]     A.D. Rast, X. Jin, F. Galluppi, L.A. Plana, C. Patterson, and S. Furber. Scalable event-driven native parallel processing: The SpiNNaker neuromimetic system. In *Proceedings of the 7th ACM International Conference on Computing Frontiers*, pages 21–30. ACM, 2010.

[RJKF09]     Alexander Rast, Xin Jin, Mukaram Khan, and Steve Furber. The deferred event model for hardware-oriented spiking neural networks. In Mario Kppen, Nikola Kasabov, and George Coghill, editors, *Advances in Neuro-Information Processing*, volume 5507 of *Lecture Notes in Computer Science*, pages 1057–1064. Springer Berlin / Heidelberg, 2009.

[RMM+08]     Patrick Rocke, Brian McGinley, John Maher, Fearghal Morgan, and Jim Harkin. Investigating the suitability of FPAAs for evolved hardware spiking neural networks. In *Proceedings of the 8th international conference on Evolvable Systems: From Biology to Hardware*, ICES '08, pages 118–129, Berlin, Heidelberg, 2008. Springer-Verlag.

[RNJ+11]     A.D. Rast, J. Navaridas, X. Jin, F. Galluppi, L.A. Plana, J. Miguel-Alonso, C. Patterson, M. Luján, and S. Furber. Managing burstiness and scalability in event-driven models on the SpiNNaker neuromimetic system. *International Journal of Parallel Programming*, pages 1–30, 2011.

[Ron96]      W. C. Rontgen. On a new kind of rays. *Science*, 3(59):227–231, 1896.

[Ror12]      Chris Rorden. *MRIcron NIfTI viewer and dcm2nii DICOM converter*. 6th May 2012. [Online; accessed 3-Jun-2012], URL: http://www.mccauslandcenter.sc.edu/mricro/.

[Ros90]      M. Rose. *Management Information Base for Network Management of TCP/IP-based Internets: MIB-II*. 1990. [Online; accessed 3-Jun-2012], URL: http://tools.ietf.org/html/rfc1158.

[Rou94]      T. Routen. Techniques for the visualisation of genetic algorithms. In *Evolutionary Computation, 1994. IEEE World Congress on Computational Intelligence., Proceedings of the First IEEE Conference on*, pages 846–851, Jun 1994.

[RPWF12]     A. D. Rast, L. A. Plana, S. R. Welbourne, and S.B. Furber. Event-Driven MLP Implementation on Neuromimetic Hardware. In *Neural Networks (IJCNN), The 2012 International Joint Conference on*. IEEE, 2012.

[RTV09]      Kenneth Rice, Tarek Taha, and Christopher Vutsinas. Scaling analysis of a neocortex inspired cognitive model on the Cray XD1. *The Journal of Supercomputing*, 47:21–43, 2009.

[SAC+11]  G. Snider, R. Amerson, D. Carter, H. Abdalla, M.S. Qureshi, J. Leveille, M. Versace, H. Ames, S. Patrick, B. Chandler, A. Gorchetchnikov, and E. Mingolla. From synapses to circuitry: Using memristive memory to explore the electronic brain. *Computer*, 44(2):21–28, Feb. 2011.

[SAL06]  Dean F. Sittig, Joan S. Ash, and Robert S. Ledley. The story behind the development of the first whole-body computerized tomography scanner as told by Robert S. Ledley. *Journal of the American Medical Informatics Association*, 13(5):465–469, 2006.

[SBLM+09]  Rebecca Smith-Bindman, Jafi Lipson, Ralph Marcus, Kwang-Pyo Kim, Mahadevappa Mahesh, Robert Gould, Amy Berrington de Gonzalez, and Diana L. Miglioretti. Radiation dose associated with common computed tomography examinations and the associated lifetime attributable risk of cancer. *Arch Intern Med*, 169(22):2078–2086, December 2009.

[SBS+95]  Thomas Sterling, Donald J Becker, Daniel Savarese, John E Dorband, Udaya A Ranawake, and Charles V Packer. BEOWULF: A Parallel Workstation For Scientific Computation. In *Parallel Processing, International Conference on*, pages 11–14, 1995.

[SFM08]  J. Schemmel, J. Fieres, and K. Meier. Wafer-scale integration of analog neural networks. In *Neural Networks, 2008. IJCNN 2008. IEEE International Joint Conference on*, pages 431–438, June 2008.

[SG98]  B. E. Swartz and E. S. Goldensohn. Timeline of the history of EEG and associated fields. *Electroencephalogr Clin Neurophysiol*, 106:173–176, Feb 1998.

[Shn96]  B. Shneiderman. The eyes have it: a task by data type taxonomy for information visualizations. In *Visual Languages, 1996. Proceedings., IEEE Symposium on*, pages 336–343, Sep 1996.

[Sim12]  Pierrick Simier. *Simple Network Management Protocol SNMP Portal*. 2012. [Online; accessed 3-Jun-2012], URL: http://snmplink.org/.

[SKC88]  T J Sejnowski, C Koch, and P S Churchland. Computational neuroscience. *Science*, 241(4871):1299–1306, September 1988.

[SLS11]  A. Salnikov, R. Levchenko, and O. Sudakov. Integrated grid environment for massive distributed computing in neuroscience. In *Intelligent Data Acquisition and Advanced Computing Systems (IDAACS), 2011 IEEE 6th International Conference on*, volume 1, pages 198–202, Sept. 2011.

[Sol12] SolarWinds. *SolarWinds: IT Management & Monitoring Software*. 2012. [Online; accessed 3-Jun-2012], URL: http://www.solarwinds.com/.

[SPF11] T. Sharp, C. Patterson, and S. Furber. Distributed configuration of massively-parallel simulation on SpiNNaker neuromorphic hardware. In *Neural Networks (IJCNN), The 2011 International Joint Conference on*, pages 1099–1105. IEEE, 2011.

[Spi12] Spiceworks Inc. *Spiceworks IT community*. 2012. [Online; accessed 3-Jun-2012], URL: http://www.spiceworks.com/.

[SSSB10] J. Somerville, L. Stuart, E. Sernagor, and R. Borisyuk. iRaster: A novel information visualization tool to explore spatiotemporal patterns in multiple spike trains. *Journal of Neuroscience Methods*, 194(1):158–171, 2010.

[Ste67] R B Stein. The Frequency of Nerve Action Potentials Generated by Applied Currents. *Proceedings of the Royal Society of London. Series B. Biological Sciences*, 167(1006):64–86, 1967.

[STE09] Terrence C Stewart, Bryan Tripp, and Chris Eliasmith. Python scripting in the Nengo simulator. *Frontiers in Neuroinformatics*, 3(0), 2009.

[SWA01] M. J. Streeter, M. O. Ward, and S. A. Alvarez. NVIS: an interactive visualization tool for neural networks. In *Society of Photo-Optical Instrumentation Engineers (SPIE) Conference Series*, volume 4302 of *Society of Photo-Optical Instrumentation Engineers (SPIE) Conference Series*, pages 234–241, May 2001.

[SyMAF00] R. Subraman yan, J. Miguel-Alonso, and J. A. B. Fortes. A scalable SNMP-based distributed monitoring system for heterogeneous network computing. In *Proc. ACM/IEEE 2000 Conference Supercomputing*, November 2000.

[TDM03] Brian J. Taylor, Marjorie A. Darrah, and Christina Moats. Verification and validation of neural networks: A sampling of research in progress. In *SPIE's 17th Annual International Symposium on Aerospace/Defense Sensing, Simulation, and Controls (AeroSense 2003)*, pages 25–27, April 2003.

[Tho90] S. J. Thorpe. *Spike arrival times: A highly efficient coding scheme for neural networks. In Parallel processing in neural systems and computers.*, pages 91–94. North-Holland Elsevier, 1990.

[Tho00] R. F. Thompson. *The Brain: A Neuroscience Primer*. W.H.Freeman and Co Ltd, ISBN: 0716732262, 3rd edition edition, 2000.

[THW02] H. Topcuoglu, S. Hariri, and Min-You Wu. Performance-effective and low-complexity task scheduling for heterogeneous computing. *Parallel and Distributed Systems, IEEE Transactions on*, 13(3):260–274, Mar 2002.

[TM05] F.-Y. Tzeng and K.-L. Ma. Opening the black box - data driven visualization of neural networks. In *Visualization, 2005. VIS 05. IEEE*, pages 383–390, Oct. 2005.

[TPPHM75] M. M. Ter-Pogossian, M. E. Phelps, E. J. Hoffman, and N. A. Mullani. A positron-emission transaxial tomograph for nuclear imaging (PETT). *Radiology*, 114:89–98, Jan 1975.

[Tuf01] Edward R. Tufte. *The Visual Display of Quantitative Information*. Graphics Press USA, 2nd edition, 2001.

[UJ06] Mat Uk and Rudolf Jaka. Framework for the interactive learning of artificial neural networks. In *Artificial Neural Networks ICANN 2006*, volume 4131 of *Lecture Notes in Computer Science*, pages 103–112. Springer Berlin / Heidelberg, 2006.

[UK 10] UK Office of the Gas and Electricity Markets (ofgem). *Electricity and Gas Supply Market Report December 2010*. 2010. [Online; accessed 3-Jun-2012], URL: http://www.ofgem.gov.uk/Pages/MoreInformation.aspx?docid=280&refer=Markets/RetMkts/ensuppro.

[Uni11] The Open University. *Open University MCT Faculty - Research in High Performance Computing*. 2011. [Online; accessed 3-Jun-2012], URL: http://mct-research.open.ac.uk/consultancyandlabservices/highperformancecomputingcluster.

[Uni12] European Union. *The Human Brain Project*. Feb 2012. [Online; accessed 3-Jun-2012], URL: http://www.humanbrainproject.eu/.

[VA05] Tim P Vogels and L F Abbott. Signal propagation and logic gating in networks of integrate-and-fire neurons. *J Neurosci*, 25(46):10786–10795, November 2005.

[VC10] M. Versace and B. Chandler. The brain of a new machine. *Spectrum, IEEE*, 47(12):30–37, December 2010.

[VMM+04] P. Volegov, A. Matlachov, J. Mosher, M. A. Espy, and R. H. Kraus. Noise-free magnetoencephalography recordings of brain function. *Phys Med Biol*, 49:2117–2128, May 2004.

[Wal91] S. Waldbusser. *Remote Network Monitoring Management Information Base*. 1991. [Online; accessed 3-Jun-2012], URL: http://tools.ietf.org/html/rfc1271.

[Wax80] Stephen G. Waxman. Determinants of conduction velocity in myelinated nerve fibers. *Muscle & Nerve*, 3(2):141–150, 1980.

[WC05]     Kevin Warwick and Daniela Cerqui. Prospects for thought communication. In *Proceedings of ETHICOMP 2005*, Linkoping, Sweden, September 2005.

[Wer74]     P. Werbos. *Beyond Regression: New Tools for Prediction and Analysis in the Behavioral Sciences*. PhD thesis, Harvard University, Cambridge, MA, 1974.

[WH88]     R. W. Williams and K. Herrup. The control of neuron number. *Annual Review of Neuroscience*, 11(1):423–453, 1988.

[Wil93]     Peter Wilke. Visualization of neural networking using neurograph. In *University Education Uses of Visualization in Scientific Computing*, pages 105–117, 1993.

[WKM11]     Thomas Williams, Colin Kelley, and Many_others. *gnuplot 4.4: an interactive plotting program*. November 2011. [Online; accessed 3-Jun-2012], URL: http://gnuplot.sourceforge.net/.

[WLK[+]11]     A. T. Woods, D. M. Lloyd, J. Kuenzel, E. Poliakoff, G. B. Dijksterhuis, and A. Thomas. Expected taste intensity affects response to sweet drinks in primary taste cortex. *Neuroreport*, 22:365–369, Jun 2011.

[WT91]     Jakub Wejchert and Gerald Tesauro. Visualizing processes in neural networks. *IBM Journal of Research and Development*, 35(1):244–253, 1991.

[Yos98]     Jeff Yoshimi. *Simbrain: A visual framework for neural network analysis and education*. 1998. [Online; accessed 3-Jun-2012], URL: http://www.simbrain.net/.

www.ingramcontent.com/pod-product-compliance
Lightning Source LLC
Chambersburg PA
CBHW081045170526
45158CB00006B/1867